Routing and Switching Essentials Lab Manual

Cisco Networking Academy

ıllıılıı CISCO™

Cisco Press
800 East 96th Street
Indianapolis, Indiana 46240

Routing and Switching Essentials Lab Manual

Cisco Networking Academy

Copyright © 2014 Cisco Systems, Inc.

Published by:
Cisco Press
800 East 96th Street
Indianapolis, IN 46240 USA

Printed in the United States of America

First Printing October 2013

Library of Congress Control Number: 2013948301

ISBN-13: 978-1-58713-320-6

ISBN-10: 1-58713-320-2

Warning and Disclaimer

This book is designed to provide information about Routing and Switching Essentials. Every effort has been made to make this book as complete and as accurate as possible, but no warranty or fitness is implied.

The information is provided on an "as is" basis. The authors, Cisco Press, and Cisco Systems, Inc. shall have neither liability nor responsibility to any person or entity with respect to any loss or damages arising from the information contained in this book or from the use of the discs or programs that may accompany it.

The opinions expressed in this book belong to the author and are not necessarily those of Cisco Systems, Inc.

Trademark Acknowledgments

All terms mentioned in this book that are known to be trademarks or service marks have been appropriately capitalized. Cisco Press or Cisco Systems, Inc., cannot attest to the accuracy of this information. Use of a term in this book should not be regarded as affecting the validity of any trademark or service mark.

This book is part of the Cisco Networking Academy® series from Cisco Press. The products in this series support and complement the Cisco Networking Academy curriculum. If you are using this book outside the Networking Academy, then you are not preparing with a Cisco trained and authorized Networking Academy provider.

For more information on the Cisco Networking Academy or to locate a Networking Academy, please visit www.cisco.com/edu.

CISCO

Feedback Information

At Cisco Press, our goal is to create in-depth technical books of the highest quality and value. Each book is crafted with care and precision, undergoing rigorous development that involves the unique expertise of members from the professional technical community.

Readers' feedback is a natural continuation of this process. If you have any comments regarding how we could improve the quality of this book, or otherwise alter it to better suit your needs, you can contact us through email at feedback@ciscopress.com. Please make sure to include the book title and ISBN in your message.

We greatly appreciate your assistance.

Publisher	**Paul Boger**
Associate Publisher	**Dave Dusthimer**
Business Operations Manager, Cisco Press	**Jan Cornelssen**
Executive Editor	**Mary Beth Ray**
Managing Editor	**Sandra Schroeder**
Project Editor	**Seth Kerney**
Editorial Assistant	**Vanessa Evans**
Cover Designer	**Mark Shirar**
Compositor	**TnT Design, Inc.**

CISCO™

Americas Headquarters	Asia Pacific Headquarters	Europe Headquarters
Cisco Systems, Inc.	Cisco Systems (USA) Pte. Ltd.	Cisco Systems International BV
San Jose, CA	Singapore	Amsterdam, The Netherlands

Cisco has more than 200 offices worldwide. Addresses, phone numbers, and fax numbers are listed on the Cisco Website at **www.cisco.com/go/offices.**

CCDE, CCENT, Cisco Eos, Cisco HealthPresence, the Cisco logo, Cisco Lumin, Cisco Nexus, Cisco StadiumVision, Cisco TelePresence, Cisco WebEx, DCE, and Welcome to the Human Network are trademarks; Changing the Way We Work, Live, Play, and Learn and Cisco Store are service marks; and Access Registrar, Aironet, AsyncOS, Bringing the Meeting To You, Catalyst, CCDA, CCDP, CCIE, CCIP, CCNA, CCNP, CCSP, CCVP, Cisco, the Cisco Certified Internetwork Expert logo, Cisco IOS, Cisco Press, Cisco Systems, Cisco Systems Capital, the Cisco Systems logo, Cisco Unity, Collaboration Without Limitation, EtherFast, EtherSwitch, Event Center, Fast Step, Follow Me Browsing, FormShare, GigaDrive, HomeLink, Internet Quotient, IOS, iPhone, iQuick Study, IronPort, the IronPort logo, LightStream, Linksys, MediaTone, MeetingPlace, MeetingPlace Chime Sound, MGX, Networkers, Networking Academy, Network Registrar, PCNow, PIX, PowerPanels, ProConnect, ScriptShare, SenderBase, SMARTnet, Spectrum Expert, StackWise, The Fastest Way to Increase Your Internet Quotient, TransPath, WebEx, and the WebEx logo are registered trademarks of Cisco Systems, Inc. and/or its affiliates in the United States and certain other countries.

All other trademarks mentioned in this document or website are the property of their respective owners. The use of the word partner does not imply a partnership relationship between Cisco and any other company. (0812R)

Contents

About This Lab Manual

Routing and Switching Essentials Lab Manual contains all the labs and class activities from the Cisco Networking Academy course of the same name. It is meant to be used within this program of study.

More Practice

If you would like more practice activities, combine your Lab Manual with the new *CCENT Practice and Study Guide* ISBN: 9781587133459

Other Related Titles

CCNA Routing and Switching Portable Command Guide ISBN: 9781587204302 (or eBook ISBN: 9780133381368)

Routing and Switching Essentials Companion Guide ISBN: 9781587133183 (or eBook ISBN: 9780133476224)

Routing and Switching Essentials Course Booklet ISBN: 9781587133190

Command Syntax Conventions

The conventions used to present command syntax in this book are the same conventions used in the IOS Command Reference. The Command Reference describes these conventions as follows:

- **Boldface** indicates commands and keywords that are entered literally as shown. In actual configuration examples and output (not general command syntax), boldface indicates commands that are manually input by the user (such as a **show** command).

- *Italic* indicates arguments for which you supply actual values.

- Vertical bars (|) separate alternative, mutually exclusive elements.

- Square brackets ([ ]) indicate an optional element.

- Braces ({ }) indicate a required choice.

- Braces within brackets ([{ }]) indicate a required choice within an optional element.

Chapter 1 — Introduction to Switched Networks

1.0.1.2 Class Activity – Sent or Received

Objectives

Describe convergence of data, voice, and video in the context of switched networks.

Scenario

Individually, or in groups (per the instructor's decision), discuss various ways hosts send and receive data, voice, and streaming video.

- Develop a matrix (table) listing network data types that can be sent and received. Provide five examples.

Your matrix table might look something like this:

Sent	Received
Client requests a web page from a web server.	Web server send web page to requesting client.

Save your work in either hard- or soft-copy format. Be prepared to discuss your matrix and statements in a class discussion.

Resources

Internet connectivity

Reflection

1. If you are receiving data, how do you think a switch assists in that process?

_____

2. If you are sending network data, how do you think a switch assists in that process?

_____

1.3.1.1 Class Activity – It's Network Access Time

Objectives

Describe features available for switches to support requirements of a small- to medium-sized business network.

Scenario

Use Packet Tracer for this activity. Work with a classmate to create two network designs to accommodate the following scenarios:

Scenario 1 – Classroom Design (LAN)

- 15 student end devices represented by 1 or 2 PCs.

- 1 instructor end device; a server is preferred.

- Device capability to stream video presentations over LAN connection. Internet connectivity is not required in this design.

Scenario 2 – Administrative Design (WAN)

- All requirements as listed in Scenario 1.

- Add access to and from a remote administrative server for video presentations and pushed updates for network application software.

Both the LAN and WAN designs should fit on to one Packet Tracer file screen. All intermediary devices should be labeled with the switch model (or name) and the router model (or name).

Save your work and be ready to justify your device decisions and layout to your instructor and the class.

Reflection

1. What are some problems that may be encountered if you receive streaming video from your instructor's server through a low-end switch?

   _____

2. How would the traffic flow be determined: multicast or broadcast – in transmission?

   _____

3. What would influence your decision on the type of switch to use for voice, streaming video and regular data these types of transmissions?

    _____

4. As you learned in the first course of the Academy, video and voice use a special TCP/IP model, transport layer protocol. What protocol is used in this layer and why is it important to voice and video streaming?

    _____

Chapter 2 — Basic Switching Concepts and Configuration

2.0.1.2 Class Activity – Stand By Me

Objective

Describe the role of unicast, broadcast, and multicast in a switched network.

Scenario

When you arrived to class today, you were given a number by your instructor to use for this introductory class activity.

Once class begins, your instructor will ask certain students with specific numbers to stand. Your job is to record the standing students' numbers for each scenario.

Scenario 1

Students with numbers **starting** with the number **5** should stand. Record the numbers of the standing students.

Scenario 2

Students with numbers **ending** in **B** should stand. Record the numbers of the standing students.

Scenario 3

Students with the number **504C** should stand. Record the number of the standing student.

At the end of this activity, divide into small groups and record answers to the Reflection questions on the PDF for this activity.

Reflection

1. Why do you think you were asked to record the students' numbers when and as requested?

   _____

2. What is the significance of the number 5 in this activity? How many people were identified with this number?

   _____

3. What is the significance of the letter C in this activity? How many people were identified with this number?

_____

4. Why did only one person stand for 504C?

    _____

5. How do you think this activity represents data travelling on local area networks?

    _____

Save your work and be prepared to share it with another student or the entire class.

2.1.1.6 Lab – Configuring Basic Switch Settings

Topology

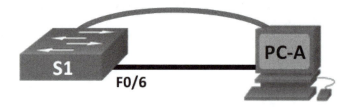

Addressing Table

Device	Interface	IP Address	Subnet Mask	Default Gateway
S1	VLAN 99	192.168.1.2	255.255.255.0	192.168.1.1
PC-A	NIC	192.168.1.10	255.255.255.0	192.168.1.1

Objectives

Part 1: Cable the Network and Verify the Default Switch Configuration

Part 2: Configure Basic Network Device Settings

- Configure basic switch settings.
- Configure the PC IP address.

Part 3: Verify and Test Network Connectivity

- Display device configuration.
- Test end-to-end connectivity with ping.
- Test remote management capabilities with Telnet.
- Save the switch running configuration file.

Part 4: Manage the MAC Address Table

- Record the MAC address of the host.
- Determine the MAC addresses that the switch has learned.
- List the **show mac address-table** command options.
- Set up a static MAC address.

Background / Scenario

Cisco switches can be configured with a special IP address known as switch virtual interface (SVI). The SVI or management address can be used for remote access to the switch to display or configure settings. If the VLAN 1 SVI is assigned an IP address, by default, all ports in VLAN 1 have access to the SVI management IP address.

In this lab, you will build a simple topology using Ethernet LAN cabling and access a Cisco switch using the console and remote access methods. You will examine default switch configurations before configuring basic switch settings. These basic switch settings include device name, interface description, local passwords, message of the day (MOTD) banner, IP addressing, setting up a static MAC address, and demonstrating the use of a management IP address for remote switch management. The topology consists of one switch and one host using only Ethernet and console ports.

Note: The switch used is a Cisco Catalyst 2960 with Cisco IOS Release 15.0(2) (lanbasek9 image). Other switches and Cisco IOS versions can be used. Depending on the model and Cisco IOS version, the commands available and output produced might vary from what is shown in the labs.

Note: Make sure that the switch has been erased and has no startup configuration. Refer to Appendix A for the procedures to initialize and reload devices.

Required Resources

- 1 Switch (Cisco 2960 with Cisco IOS Release 15.0(2) lanbasek9 image or comparable)
- 1 PC (Windows 7, Vista, or XP with terminal emulation program, such as Tera Term, and Telnet capability)
- Console cable to configure the Cisco IOS device via the console port
- Ethernet cable as shown in the topology

Part 1:   Cable the Network and Verify the Default Switch Configuration

In Part 1, you will set up the network topology and verify default switch settings.

Step 1: Cable the network as shown in the topology.

a. Cable the console connection as shown in the topology. Do not connect the PC-A Ethernet cable at this time.

   Note: If you are using Netlab, you can shut down F0/6 on S1 which has the same effect as not connecting PC-A to S1.

b. Create a console connection to the switch from PC-A using Tera Term or other terminal emulation program.

   Why must you use a console connection to initially configure the switch? Why is it not possible to connect to the switch via Telnet or SSH?

_____

Step 2: Verify the default switch configuration.

In this step, you will examine the default switch settings, such as current switch configuration, IOS information, interface properties, VLAN information, and flash memory.

You can access all the switch IOS commands in privileged EXEC mode. Access to privileged EXEC mode should be restricted by password protection to prevent unauthorized use because it provides direct access to global configuration mode and commands used to configure operating parameters. You will set passwords later in this lab.

The privileged EXEC mode command set includes those commands contained in user EXEC mode, as well as the **configure** command through which access to the remaining command modes is gained. Use the **enable** command to enter privileged EXEC mode.

a. Assuming the switch had no configuration file stored in nonvolatile random-access memory (NVRAM), you will be at the user EXEC mode prompt on the switch with a prompt of Switch>. Use the **enable** command to enter privileged EXEC mode.

```
Switch> enable
Switch#
```

Notice that the prompt changed in the configuration to reflect privileged EXEC mode.

Verify a clean configuration file with the **show running-config** privileged EXEC mode command. If a configuration file was previously saved, it must be removed. Depending on switch model and IOS version, your configuration may look slightly different. However, there should be no configured passwords or IP address. If your switch does not have a default configuration, erase and reload the switch.

Note: Appendix A details the steps to initialize and reload the devices.

b. Examine the current running configuration file.

```
Switch# show running-config
```

How many FastEthernet interfaces does a 2960 switch have? _____

How many Gigabit Ethernet interfaces does a 2960 switch have? _____

What is the range of values shown for the vty lines? _____

c. Examine the startup configuration file in NVRAM.

```
Switch# show startup-config
startup-config is not present
```

Why does this message appear? _____

d. Examine the characteristics of the SVI for VLAN 1.

```
Switch# show interface vlan1
```

Is there an IP address assigned to VLAN 1? _____

What is the MAC address of this SVI? Answers will vary. _____

Is this interface up?

_____

e. Examine the IP properties of the SVI VLAN 1.

```
Switch# show ip interface vlan1
```

What output do you see?

_____

_____

f. Connect PC-A Ethernet cable to port 6 on the switch and examine the IP properties of the SVI VLAN 1. Allow time for the switch and PC to negotiate duplex and speed parameters.

Note: If you are using Netlab, enable interface F0/6 on S1.

Switch# **show ip interface vlan1**

What output do you see?

_____

_____

g. Examine the Cisco IOS version information of the switch.

Switch# **show version**

What is the Cisco IOS version that the switch is running? _____

What is the system image filename? _____

What is the base MAC address of this switch? Answers will vary. _____

h. Examine the default properties of the FastEthernet interface used by PC-A.

Switch# **show interface f0/6**

Is the interface up or down? _____

What event would make an interface go up? _____

What is the MAC address of the interface? _____

What is the speed and duplex setting of the interface? _____

i. Examine the default VLAN settings of the switch.

Switch# **show vlan**

What is the default name of VLAN 1? _____

Which ports are in this VLAN? _____

Is VLAN 1 active? _____

What type of VLAN is the default VLAN? _____

j. Examine flash memory.

Issue one of the following commands to examine the contents of the flash directory.

```
Switch# show flash
```

```
Switch# dir flash:
```

Files have a file extension, such as .bin, at the end of the filename. Directories do not have a file extension.

What is the filename of the Cisco IOS image? _____

Part 2:   Configure Basic Network Device Settings

In Part 2, you configure basic settings for the switch and PC.

Step 1:   Configure basic switch settings including hostname, local passwords, MOTD banner, management address, and Telnet access.

In this step, you will configure the PC and basic switch settings, such as hostname and an IP address for the switch management SVI. Assigning an IP address on the switch is only the first step. As the network administrator, you must specify how the switch is managed. Telnet and SSH are the two most common management methods. However, Telnet is not a secure protocol. All information flowing between the two devices is sent in plain text. Passwords and other sensitive information can be easily looked at if captured by a packet sniffer.

a. Assuming the switch had no configuration file stored in NVRAM, verify you are at privileged EXEC mode. Enter **enable** if the prompt has changed back to Switch>.

```
Switch> enable
Switch#
```

b. Enter global configuration mode.

```
Switch# configure terminal
Enter configuration commands, one per line. End with CNTL/Z.
Switch(config)#
```

The prompt changed again to reflect global configuration mode.

c. Assign the switch hostname.

```
Switch(config)# hostname S1
S1(config)#
```

d. Configure password encryption.

```
S1(config)# service password-encryption
S1(config)#
```

e. Assign **class** as the secret password for privileged EXEC mode access.

```
S1(config)# enable secret class
S1(config)#
```

f.   Prevent unwanted DNS lookups.

```
S1(config)# no ip domain-lookup
S1(config)#
```

g.   Configure a MOTD banner.

```
S1(config)# banner motd #
Enter Text message.  End with the character '#'.
Unauthorized access is strictly prohibited. #
```

h.   Verify your access settings by moving between modes.

```
S1(config)# exit
S1#
*Mar  1 00:19:19.490: %SYS-5-CONFIG_I: Configured from console by console
S1# exit

S1 con0 is now available




Press RETURN to get started.



Unauthorized access is strictly prohibited.
S1>
```

Which shortcut keys are used to go directly from global configuration mode to privileged EXEC mode?

_____

i.   Go back to privileged EXEC mode from user EXEC mode. Enter **class** as the password when prompted.

```
S1> enable
Password:
S1#
```

Note: The password does not display when entering.

j.   Enter global configuration mode to set the SVI IP address of the switch. This allows remote management of the switch.

Before you can manage S1 remotely from PC-A, you must assign the switch an IP address. The default configuration on the switch is to have the management of the switch controlled through VLAN 1. However, a best practice for basic switch configuration is to change the management VLAN to a VLAN other than VLAN 1.

For management purposes, use VLAN 99. The selection of VLAN 99 is arbitrary and in no way implies that you should always use VLAN 99.

First, create the new VLAN 99 on the switch. Then set the IP address of the switch to 192.168.1.2 with a subnet mask of 255.255.255.0 on the internal virtual interface VLAN 99.

```
S1# configure terminal
S1(config)# vlan 99
S1(config-vlan)# exit
S1(config)# interface vlan99
%LINEPROTO-5-UPDOWN: Line protocol on Interface Vlan99, changed state to down
S1(config-if)# ip address 192.168.1.2 255.255.255.0
S1(config-if)# no shutdown
S1(config-if)# exit
S1(config)#
```

Notice that the VLAN 99 interface is in the down state even though you entered the **no shutdown** command. The interface is currently down because no switch ports are assigned to VLAN 99.

k. Assign all user ports to VLAN 99.

```
S1(config)# interface range f0/1 - 24,g0/1 - 2
S1(config-if-range)# switchport access vlan 99
S1(config-if-range)# exit
S1(config)#
%LINEPROTO-5-UPDOWN: Line protocol on Interface Vlan1, changed state to down
%LINEPROTO-5-UPDOWN: Line protocol on Interface Vlan99, changed state to up
```

To establish connectivity between the host and the switch, the ports used by the host must be in the same VLAN as the switch. Notice in the above output that the VLAN 1 interface goes down because none of the ports are assigned to VLAN 1. After a few seconds, VLAN 99 comes up because at least one active port (F0/6 with PC-A attached) is now assigned to VLAN 99.

l. Issue **show vlan brief** command to verify that all the user ports are in VLAN 99.

```
S1# show vlan brief


VLAN Name                             Status    Ports
---- -------------------------------- --------- -------------------------------
1    default                          active
99   VLAN0099                         active    Fa0/1, Fa0/2, Fa0/3, Fa0/4
                                                Fa0/5, Fa0/6, Fa0/7, Fa0/8
                                                Fa0/9, Fa0/10, Fa0/11, Fa0/12
                                                Fa0/13, Fa0/14, Fa0/15, Fa0/16
                                                Fa0/17, Fa0/18, Fa0/19, Fa0/20
                                                Fa0/21, Fa0/22, Fa0/23, Fa0/24
                                                Gi0/1, Gi0/2
1002 fddi-default                     act/unsup
1003 token-ring-default               act/unsup
1004 fddinet-default                  act/unsup
1005 trnet-default                    act/unsup
```

m. Configure the IP default gateway for S1. If no default gateway is set, the switch cannot be managed from a remote network that is more than one router away. It does respond to pings from a remote network. Although this activity does not include an external IP gateway, assume that you will eventually connect the LAN to a router for external access. Assuming that the LAN interface on the router is 192.168.1.1, set the default gateway for the switch.

```
S1(config)# ip default-gateway 192.168.1.1

S1(config)#
```

n. Console port access should also be restricted. The default configuration is to allow all console connections with no password needed. To prevent console messages from interrupting commands, use the **logging synchronous** option.

```
S1(config)# line con 0

S1(config-line)# password cisco

S1(config-line)# login

S1(config-line)# logging synchronous

S1(config-line)# exit

S1(config)#
```

o. Configure the virtual terminal (vty) lines for the switch to allow Telnet access. If you do not configure a vty password, you are unable to telnet to the switch.

```
S1(config)# line vty 0 15

S1(config-line)# password cisco

S1(config-line)# login

S1(config-line)# end

S1#

*Mar  1 00:06:11.590: %SYS-5-CONFIG_I: Configured from console by console
```

Why is the **login** command required? _____

Step 2: Configure an IP address on PC-A.

Assign the IP address and subnet mask to the PC as shown in the Addressing Table. An abbreviated version of the procedure is described here. A default gateway is not required for this topology; however, you can enter **192.168.1.1** to simulate a router attached to S1.

1) Click the Windows **Start** icon > **Control Panel**.

2) Click **View By:** and choose **Small icons**.

3) Choose **Network and Sharing Center** > **Change adapter settings**.

4) Select **Local Area Network Connection,** right click and choose **Properties**.

5) Choose **Internet Protocol Version 4 (TCP/IPv4)** > **Properties**.

6) Click the **Use the following IP address** radio button and enter the IP address and subnet mask.

Part 3:   Verify and Test Network Connectivity

In Part 3, you will verify and document the switch configuration, test end-to-end connectivity between PC-A and S1, and test the switch's remote management capability.

Step 1:   Display the switch configuration.

From your console connection on PC-A, display and verify your switch configuration. The **show run** command displays the entire running configuration, one page at a time. Use the spacebar to advance paging.

a.   A sample configuration displays here. The settings you configured are highlighted in yellow. The other configuration settings are IOS defaults.

```
S1# show run
Building configuration...

Current configuration : 2206 bytes
!
version 15.0
no service pad
service timestamps debug datetime msec
service timestamps log datetime msec
service password-encryption
!
hostname S1
!
boot-start-marker
boot-end-marker
!
enable secret 4 06YFDUHH61wAE/kLkDq9BGho1QM5EnRtoyr8cHAUg.2
!
no aaa new-model
system mtu routing 1500
!
!
no ip domain-lookup
!
<output omitted>
!
interface FastEthernet0/24
 switchport access vlan 99
!
interface GigabitEthernet0/1
!
interface GigabitEthernet0/2
```

```
!
interface Vlan1
 no ip address
 no ip route-cache
!
interface Vlan99
 ip address 192.168.1.2 255.255.255.0
 no ip route-cache
!
ip default-gateway 192.168.1.1
ip http server
ip http secure-server
!
banner motd ^C
Unauthorized access is strictly prohibited. ^C
!
line con 0
 password 7 104D000A0618
 logging synchronous
 login
line vty 0 4
 password 7 14141B180F0B
 login
line vty 5 15
 password 7 14141B180F0B
 login
!
end

S1#
```

b. Verify the management VLAN 99 settings.

```
S1# show interface vlan 99
Vlan99 is up, line protocol is up
  Hardware is EtherSVI, address is 0cd9.96e2.3d41 (bia 0cd9.96e2.3d41)
  Internet address is 192.168.1.2/24
  MTU 1500 bytes, BW 1000000 Kbit, DLY 10 usec,
     reliability 255/255, txload 1/255, rxload 1/255
  Encapsulation ARPA, loopback not set
  ARP type: ARPA, ARP Timeout 04:00:00
  Last input 00:00:06, output 00:08:45, output hang never
  Last clearing of "show interface" counters never
```

```
Input queue: 0/75/0/0 (size/max/drops/flushes); Total output drops: 0
Queueing strategy: fifo
Output queue: 0/40 (size/max)
5 minute input rate 0 bits/sec, 0 packets/sec
5 minute output rate 0 bits/sec, 0 packets/sec
   175 packets input, 22989 bytes, 0 no buffer
   Received 0 broadcasts (0 IP multicast)
   0 runts, 0 giants, 0 throttles
   0 input errors, 0 CRC, 0 frame, 0 overrun, 0 ignored
   1 packets output, 64 bytes, 0 underruns
   0 output errors, 0 interface resets
   0 output buffer failures, 0 output buffers swapped out
```

What is the bandwidth on this interface? _____

What is the VLAN 99 state? _____

What is the line protocol state? _____

Step 2:   **Test end-to-end connectivity with ping.**

a.   From the command prompt on PC-A, ping your own PC-A address first.

```
C:\Users\User1> ping 192.168.1.10
```

b.   From the command prompt on PC-A, ping the SVI management address of S1.

```
C:\Users\User1> ping 192.168.1.2
```

Because PC-A needs to resolve the MAC address of S1 through ARP, the first packet may time out. If ping results continue to be unsuccessful, troubleshoot the basic device configurations. You should check both the physical cabling and logical addressing if necessary.

Step 3:   **Test and verify remote management of S1.**

You will now use Telnet to remotely access the switch. In this lab, PC-A and S1 reside side by side. In a production network, the switch could be in a wiring closet on the top floor while your management PC is located on the ground floor. In this step, you will use Telnet to remotely access switch S1 using its SVI management address. Telnet is not a secure protocol; however, you will use it to test remote access. With Telnet, all information, including passwords and commands, are sent across the session in plain text. In subsequent labs, you will use SSH to remotely access network devices.

Note: If you are using Windows 7, the administrator may need to enable the Telnet protocol. To install the Telnet client, open a cmd window and type **pkgmgr /iu:"TelnetClient"**. An example is shown below.

```
C:\Users\User1> pkgmgr /iu:"TelnetClient"
```

a. With the cmd window still open on PC-A, issue a Telnet command to connect to S1 via the SVI management address. The password is **cisco**.

```
C:\Users\User1> telnet 192.168.1.2
```

b. After entering the password **cisco**, you will be at the user EXEC mode prompt. Access privileged EXEC mode.

c. Type **exit** to end the Telnet session.

Step 4: Save the switch running configuration file.

Save the configuration.

```
S1# copy running-config startup-config

Destination filename [startup-config]? [Enter]

Building configuration...

[OK]

S1#
```

Part 4: Manage the MAC Address Table

In Part 4, you will determine the MAC address that the switch has learned, set up a static MAC address on one interface of the switch, and then remove the static MAC address from that interface.

Step 1: Record the MAC address of the host.

From a command prompt on PC-A, issue **ipconfig /all** command to determine and record the Layer 2 (physical) addresses of the PC NIC.

_____

Step 2: Determine the MAC addresses that the switch has learned.

Display the MAC addresses using the **show mac address-table** command.

```
S1# show mac address-table
```

How many dynamic addresses are there? _____

How many MAC addresses are there in total? _____

Does the dynamic MAC address match the PC-A MAC address? _____

Step 3: List the show mac address-table options.

a. Display the MAC address table options.

```
S1# show mac address-table ?
```

How many options are available for the **show mac address-table** command? _____

b. Issue the **show mac address-table dynamic** command to display only the MAC addresses that were learned dynamically.

```
S1# show mac address-table dynamic
```

How many dynamic addresses are there? _____

c. View the MAC address entry for PC-A. The MAC address formatting for the command is xxxx.xxxx.xxxx.

```
S1# show mac address-table address <PC-A MAC here>
```

Step 4: Set up a static MAC address.

a. Clear the MAC address table.

To remove the existing MAC addresses, use the **clear mac address-table** command from privileged EXEC mode.

```
S1# clear mac address-table dynamic
```

b. Verify that the MAC address table was cleared.

```
S1# show mac address-table
```

How many static MAC addresses are there? _____

How many dynamic addresses are there? _____

c. Examine the MAC table again.

More than likely, an application running on your PC has already sent a frame out the NIC to S1. Look at the MAC address table again in privileged EXEC mode to see if S1 has relearned the MAC address for PC-A.

```
S1# show mac address-table
```

How many dynamic addresses are there? _____

Why did this change from the last display? _____

If S1 has not yet relearned the MAC address for PC-A, ping the VLAN 99 IP address of the switch from PC-A, and then repeat the **show mac address-table** command.

d. Set up a static MAC address.

To specify which ports a host can connect to, one option is to create a static mapping of the host MAC address to a port.

Set up a static MAC address on F0/6 using the address that was recorded for PC-A in Part 4, Step 1. The MAC address 0050.56BE.6C89 is used as an example only. You must use the MAC address of your PC-A, which is different than the one given here as an example.

```
S1(config)# mac address-table static 0050.56BE.6C89 vlan 99 interface
fastethernet 0/6
```

e. Verify the MAC address table entries.

S1# **show mac address-table**

How many total MAC addresses are there? _____

How many static addresses are there? _____

f. Remove the static MAC entry. Enter global configuration mode and remove the command by putting a **no** in front of the command string.

Note: The MAC address 0050.56BE.6C89 is used in the example only. Use the MAC address for your PC-A.

S1(config)# **no mac address-table static 0050.56BE.6C89 vlan 99 interface fastethernet 0/6**

g. Verify that the static MAC address has been cleared.

S1# **show mac address-table**

How many total static MAC addresses are there? _____

Reflection

1. Why should you configure the vty lines for the switch?

   _____

2. Why change the default VLAN 1 to a different VLAN number?

   _____

3. How can you prevent passwords from being sent in plain text?

   _____

4. Why configure a static MAC address on a port interface?

   _____

Appendix A: Initializing and Reloading a Router and Switch

Step 1: **Initialize and reload the router.**

a. Console into the router and enable privileged EXEC mode.

Router> **enable**

Router#

b. Enter the **erase startup-config** command to remove the startup configuration from NVRAM.

```
Router# erase startup-config

Erasing the nvram filesystem will remove all configuration files! Continue? [confirm]

[OK]

Erase of nvram: complete

Router#
```

c. Issue the **reload** command to remove an old configuration from memory. When prompted to Proceed with reload?, press Enter. (Pressing any other key aborts the reload.)

```
Router# reload

Proceed with reload? [confirm]

*Nov 29 18:28:09.923: %SYS-5-RELOAD: Reload requested by console. Reload Reason: Re-
load Command.
```

Note: You may receive a prompt asking to save the running configuration prior to reloading the router. Respond by typing **no** and press Enter.

```
System configuration has been modified. Save? [yes/no]: no
```

d. After the router reloads, you are prompted to enter the initial configuration dialog. Enter **no** and press Enter.

```
Would you like to enter the initial configuration dialog? [yes/no]: no
```

e. Another prompt asks to terminate autoinstall. Respond by typing **yes** press Enter.

```
Would you like to terminate autoinstall? [yes]: yes
```

Step 2: **Initialize and reload the switch.**

a. Console into the switch and enter privileged EXEC mode.

```
Switch> enable
Switch#
```

b. Use the **show flash** command to determine if any VLANs have been created on the switch.

```
Switch# show flash

Directory of flash:/

    2  -rwx        1919   Mar 1 1993 00:06:33 +00:00   private-config.text
    3  -rwx        1632   Mar 1 1993 00:06:33 +00:00   config.text
    4  -rwx       13336   Mar 1 1993 00:06:33 +00:00   multiple-fs
    5  -rwx    11607161   Mar 1 1993 02:37:06 +00:00   c2960-lanbasek9-mz.150-2.SE.bin
    6  -rwx         616   Mar 1 1993 00:07:13 +00:00   vlan.dat

32514048 bytes total (20886528 bytes free)
Switch#
```

c. If the **vlan.dat** file was found in flash, then delete this file.

```
Switch# delete vlan.dat

Delete filename [vlan.dat]?
```

d. You are prompted to verify the filename. If you have entered the name correctly, press Enter; otherwise, you can change the filename.

e. You are prompted to confirm to delete this file. Press Enter to confirm.

```
Delete flash:/vlan.dat? [confirm]
Switch#
```

f. Use the **erase startup-config** command to erase the startup configuration file from NVRAM. You are prompted to remove the configuration file. Press Enter to confirm.

```
Switch# erase startup-config
Erasing the nvram filesystem will remove all configuration files! Continue? [confirm]
[OK]
Erase of nvram: complete
Switch#
```

g. Reload the switch to remove any old configuration information from memory. You will then receive a prompt to confirm to reload the switch. Press Enter to proceed.

```
Switch# reload
Proceed with reload? [confirm]
```

Note: You may receive a prompt to save the running configuration prior to reloading the switch. Respond by typing **no** and press Enter.

```
System configuration has been modified. Save? [yes/no]: no
```

h. After the switch reloads, you should see a prompt to enter the initial configuration dialog. Respond by entering **no** at the prompt and press Enter.

```
Would you like to enter the initial configuration dialog? [yes/no]: no
Switch>
```

2.2.4.11 Lab – Configuring Switch Security Features

Topology

Addressing Table

Device	Interface	IP Address	Subnet Mask	Default Gateway
R1	G0/1	172.16.99.1	255.255.255.0	N/A
S1	VLAN 99	172.16.99.11	255.255.255.0	172.16.99.1
PC-A	NIC	172.16.99.3	255.255.255.0	172.16.99.1

Objectives

Part 1: Set Up the Topology and Initialize Devices

Part 2: Configure Basic Device Settings and Verify Connectivity

Part 3: Configure and Verify SSH Access on S1

- Configure SSH access.
- Modify SSH parameters.
- Verify the SSH configuration.

Part 4: Configure and Verify Security Features on S1

- Configure and verify general security features.
- Configure and verify port security.

Background / Scenario

It is quite common to lock down access and install good security features on PCs and servers. It is important that your network infrastructure devices, such as switches and routers, are also configured with security features.

In this lab, you will follow some best practices for configuring security features on LAN switches. You will only allow SSH and secure HTTPS sessions. You will also configure and verify port security to lock out any device with a MAC address not recognized by the switch.

Note: The router used with CCNA hands-on labs is a Cisco 1941 Integrated Services Router (ISR) with Cisco IOS Release 15.2(4)M3 (universalk9 image). The switch used is a Cisco Catalyst 2960 with Cisco IOS Release 15.0(2) (lanbasek9 image). Other routers, switches, and Cisco IOS versions can be used. Depending on the model and Cisco IOS version, the commands available and output produced might vary from what is shown in the labs. Refer to the Router Interface Summary Table at the end of this lab for the correct interface identifiers.

Note: Make sure that the router and switch have been erased and have no startup configurations. If you are unsure, contact your instructor or refer to the previous lab for the procedures to initialize and reload devices.

Required Resources

- 1 Router (Cisco 1941 with Cisco IOS Release 15.2(4)M3 universal image or comparable)
- 1 Switch (Cisco 2960 with Cisco IOS Release 15.0(2) lanbasek9 image or comparable)
- 1 PC (Windows 7, Vista, or XP with terminal emulation program, such as Tera Term)
- Console cables to configure the Cisco IOS devices via the console ports
- Ethernet cables as shown in the topology

Part 1:   Set Up the Topology and Initialize Devices

In Part 1, you will set up the network topology and clear any configurations if necessary.

Step 1:   Cable the network as shown in the topology.

Step 2:   Initialize and reload the router and switch.

If configuration files were previously saved on the router or switch, initialize and reload these devices back to their basic configurations.

Part 2:   Configure Basic Device Settings and Verify Connectivity

In Part 2, you configure basic settings on the router, switch, and PC. Refer to the Topology and Addressing Table at the beginning of this lab for device names and address information.

Step 1:   Configure an IP address on PC-A.

Step 2:   Configure basic settings on R1.

a.   Configure the device name.

b.   Disable DNS lookup.

c.   Configure interface IP address as shown in the Addressing Table.

d.   Assign **class** as the privileged EXEC mode password.

e.   Assign **cisco** as the console and vty password and enable login.

f.   Encrypt plain text passwords.

g.   Save the running configuration to startup configuration.

Step 3:   **Configure basic settings on S1.**

A good security practice is to assign the management IP address of the switch to a VLAN other than VLAN 1 (or any other data VLAN with end users). In this step, you will create VLAN 99 on the switch and assign it an IP address.

a.   Configure the device name.

b.   Disable DNS lookup.

c.   Assign **class** as the privileged EXEC mode password.

d.   Assign **cisco** as the console and vty password and then enable login.

e.   Configure a default gateway for S1 using the IP address of R1.

f.   Encrypt plain text passwords.

g.   Save the running configuration to startup configuration.

h.   Create VLAN 99 on the switch and name it **Management**.

```
S1(config)# vlan 99
S1(config-vlan)# name Management
S1(config-vlan)# exit
S1(config)#
```

i.   Configure the VLAN 99 management interface IP address, as shown in the Addressing Table, and enable the interface.

```
S1(config)# interface vlan 99
S1(config-if)# ip address 172.16.99.11 255.255.255.0
S1(config-if)# no shutdown
S1(config-if)# end
S1#
```

j.   Issue the **show vlan** command on S1. What is the status of VLAN 99? _____

k.   Issue the **show ip interface brief** command on S1. What is the status and protocol for management interface VLAN 99?

_____

Why is the protocol down, even though you issued the **no shutdown** command for interface VLAN 99?

_____

l.   Assign ports F0/5 and F0/6 to VLAN 99 on the switch.

```
S1# config t
S1(config)# interface f0/5
S1(config-if)# switchport mode access
S1(config-if)# switchport access vlan 99
S1(config-if)# interface f0/6
S1(config-if)# switchport mode access
S1(config-if)# switchport access vlan 99
S1(config-if)# end
```

m.   Issue the **show ip interface brief** command on S1. What is the status and protocol showing for interface VLAN 99? _____

Note: There may be a delay while the port states converge.

Step 4:   Verify connectivity between devices.

a.   From PC-A, ping the default gateway address on R1. Were your pings successful? _____

b.   From PC-A, ping the management address of S1. Were your pings successful? _____

c.   From S1, ping the default gateway address on R1. Were your pings successful? _____

d.   From PC-A, open a web browser and go to http://172.16.99.11. If it prompts you for a username and password, leave the username blank and use **class** for the password. If it prompts for secured connection, answer **No**. Were you able to access the web interface on S1? _____

e.   Close the browser session on PC-A.

Note: The non-secure web interface (HTTP server) on a Cisco 2960 switch is enabled by default. A common security measure is to disable this service, as described in Part 4.

Part 3:   **Configure and Verify SSH Access on S1**

Step 1:   **Configure SSH access on S1.**

a.   Enable SSH on S1. From global configuration mode, create a domain name of **CCNA-Lab.com**.

```
S1(config)# ip domain-name CCNA-Lab.com
```

b.   Create a local user database entry for use when connecting to the switch via SSH. The user should have administrative level access.

Note: The password used here is NOT a strong password. It is merely being used for lab purposes.

```
S1(config)# username admin privilege 15 secret sshadmin
```

c.   Configure the transport input for the vty lines to allow SSH connections only, and use the local database for authentication.

```
S1(config)# line vty 0 15
S1(config-line)# transport input ssh
S1(config-line)# login local
S1(config-line)# exit
```

d.   Generate an RSA crypto key using a modulus of 1024 bits.

```
S1(config)# crypto key generate rsa modulus 1024
The name for the keys will be: S1.CCNA-Lab.com

% The key modulus size is 1024 bits
% Generating 1024 bit RSA keys, keys will be non-exportable...
[OK] (elapsed time was 3 seconds)

S1(config)#
S1(config)# end
```

e.   Verify the SSH configuration and answer the questions below.

```
S1# show ip ssh
```

What version of SSH is the switch using? _____

How many authentication attempts does SSH allow? _____

What is the default timeout setting for SSH? _____

Step 2:   Modify the SSH configuration on S1.

Modify the default SSH configuration.

```
S1# config t
S1(config)# ip ssh time-out 75
S1(config)# ip ssh authentication-retries 2
```

How many authentication attempts does SSH allow? _____

What is the timeout setting for SSH? _____

Step 3:   **Verify the SSH configuration on S1.**

a. Using SSH client software on PC-A (such as Tera Term), open an SSH connection to S1. If you receive a message on your SSH client regarding the host key, accept it. Log in with **admin** for username and **cisco** for the password.

Was the connection successful? _____

What prompt was displayed on S1? Why?

b. Type **exit** to end the SSH session on S1.

Part 4:   **Configure and Verify Security Features on S1**

In Part 4, you will shut down unused ports, turn off certain services running on the switch, and configure port security based on MAC addresses. Switches can be subject to MAC address table overflow attacks, MAC spoofing attacks, and unauthorized connections to switch ports. You will configure port security to limit the number of MAC addresses that can be learned on a switch port and disable the port if that number is exceeded.

Step 1:   **Configure general security features on S1.**

a. Configure a message of the day (MOTD) banner on S1 with an appropriate security warning message.

b. Issue a **show ip interface brief** command on S1. What physical ports are up?

_____

c. Shut down all unused physical ports on the switch. Use the **interface range** command.

```
S1(config)# interface range f0/1 - 4
S1(config-if-range)# shutdown
S1(config-if-range)# interface range f0/7 - 24
S1(config-if-range)# shutdown
S1(config-if-range)# interface range g0/1 - 2
S1(config-if-range)# shutdown
S1(config-if-range)# end
S1#
```

d. Issue the **show ip interface brief** command on S1. What is the status of ports F0/1 to F0/4?

_____

e. Issue the **show ip http server status** command.

What is the HTTP server status? _____

What server port is it using? _____

What is the HTTP secure server status? _____

What secure server port is it using? _____

f.   HTTP sessions send everything in plain text. You will disable the HTTP service running on S1.

```
S1(config)# no ip http server
```

g.   From PC-A, open a web browser session to http://172.16.99.11. What was your result?

_____

h.   From PC-A, open a secure web browser session at https://172.16.99.11. Accept the certificate. Log in with no username and a password of **class**. What was your result?

_____

i.   Close the web session on PC-A.

Step 2:   Configure and verify port security on S1.

a.   Record the R1 G0/1 MAC address. From the R1 CLI, use the **show interface g0/1** command and record the MAC address of the interface.

```
R1# show interface g0/1

GigabitEthernet0/1 is up, line protocol is up

  Hardware is CN Gigabit Ethernet, address is 30f7.0da3.1821 (bia
3047.0da3.1821)
```

What is the MAC address of the R1 G0/1 interface?

_____

b.   From the S1 CLI, issue a **show mac address-table** command from privileged EXEC mode. Find the dynamic entries for ports F0/5 and F0/6. Record them below.

F0/5 MAC address: _____

F0/6 MAC address: _____

c.   Configure basic port security.

Note: This procedure would normally be performed on all access ports on the switch. F0/5 is shown here as an example.

1)   From the S1 CLI, enter interface configuration mode for the port that connects to R1.

```
S1(config)# interface f0/5
```

2) Shut down the port.

```
S1(config-if)# shutdown
```

3) Enable port security on F0/5.

```
S1(config-if)# switchport port-security
```

Note: Entering the **switchport port-security** command sets the maximum MAC addresses to 1 and the violation action to shutdown. The **switchport port-security maximum** and **switchport port-security violation** commands can be used to change the default behavior.

4) Configure a static entry for the MAC address of R1 G0/1 interface recorded in Step 2a.

```
S1(config-if)# switchport port-security mac-address xxxx.xxxx.xxxx
```

(xxxx.xxxx.xxxx is the actual MAC address of the router G0/1 interface)

Note: Optionally, you can use the **switchport port-security mac-address sticky** command to add all the secure MAC addresses that are dynamically learned on a port (up to the maximum set) to the switch running configuration.

5) Enable the switch port.

```
S1(config-if)# no shutdown
S1(config-if)# end
```

d. Verify port security on S1 F0/5 by issuing a **show port-security interface** command.

```
S1# show port-security interface f0/5
Port Security              : Enabled
Port Status                : Secure-up
Violation Mode             : Shutdown
Aging Time                 : 0 mins
Aging Type                 : Absolute
SecureStatic Address Aging : Disabled
Maximum MAC Addresses      : 1
Total MAC Addresses        : 1
Configured MAC Addresses   : 1
Sticky MAC Addresses       : 0
Last Source Address:Vlan   : 0000.0000.0000:0
Security Violation Count   : 0
```

What is the port status of F0/5?

_____

e. From R1 command prompt, ping PC-A to verify connectivity.

```
R1# ping 172.16.99.3
```

f.   You will now violate security by changing the MAC address on the router interface. Enter interface configuration mode for G0/1 and shut it down.

```
R1# config t

R1(config)# interface g0/1

R1(config-if)# shutdown
```

g.   Configure a new MAC address for the interface, using **aaaa.bbbb.cccc** as the address.

```
R1(config-if)# mac-address aaaa.bbbb.cccc
```

h.   If possible, have a console connection open on S1 at the same time that you do this step. You will see various messages displayed on the console connection to S1 indicating a security violation. Enable the G0/1 interface on R1.

```
R1(config-if)# no shutdown
```

i.   From R1 privilege EXEC mode, ping PC-A. Was the ping successful? Why or why not?

_____

j.   On the switch, verify port security with the following commands shown below.

```
S1# show port-security
Secure Port MaxSecureAddr CurrentAddr SecurityViolation Security Action
            (Count)       (Count)     (Count)
---------------------------------------------------------------
     Fa0/5          1             1                 1         Shutdown
---------------------------------------------------------------
Total Addresses in System (excluding one mac per port)    :0
Max Addresses limit in System (excluding one mac per port) :8192


S1# show port-security interface f0/5
Port Security              : Enabled
Port Status                : Secure-shutdown
Violation Mode             : Shutdown
Aging Time                 : 0 mins
Aging Type                 : Absolute
SecureStatic Address Aging : Disabled
Maximum MAC Addresses      : 1
Total MAC Addresses        : 1
Configured MAC Addresses   : 1
Sticky MAC Addresses       : 0
Last Source Address:Vlan   : aaaa.bbbb.cccc:99
Security Violation Count   : 1


S1# show interface f0/5
```

```
FastEthernet0/5 is down, line protocol is down (err-disabled)
   Hardware is Fast Ethernet, address is 0cd9.96e2.3d05 (bia 0cd9.96e2.3d05)
   MTU 1500 bytes, BW 10000 Kbit/sec, DLY 1000 usec,
      reliability 255/255, txload 1/255, rxload 1/255
<output omitted>
```

```
S1# show port-security address
                 Secure Mac Address Table
-------------------------------------------------------------------
Vlan    Mac Address        Type              Ports     Remaining Age
                                                        (mins)
----    -----------        ----              -----     -------------
  99    30f7.0da3.1821     SecureConfigured  Fa0/5       -
-------------------------------------------------------------------
Total Addresses in System (excluding one mac per port)    :0
Max Addresses limit in System (excluding one mac per port) :8192
```

k. On the router, shut down the G0/1 interface, remove the hard-coded MAC address from the router, and re-enable the G0/1 interface.

```
R1(config-if)# shutdown
R1(config-if)# no mac-address aaaa.bbbb.cccc
R1(config-if)# no shutdown
R1(config-if)# end
```

l. From R1, ping PC-A again at 172.16.99.3. Was the ping successful? _____

m. Issue the **show interface f0/5** command to determine the cause of ping failure. Record your findings.

_____

n. Clear the S1 F0/5 error disabled status.

```
S1# config t
S1(config)# interface f0/5
S1(config-if)# shutdown
S1(config-if)# no shutdown
```

Note: There may be a delay while the port states converge.

o. Issue the **show interface f0/5** command on S1 to verify F0/5 is no longer in error disabled mode.

```
S1# show interface f0/5
FastEthernet0/5 is up, line protocol is up (connected)
   Hardware is Fast Ethernet, address is 0023.5d59.9185 (bia 0023.5d59.9185)
   MTU 1500 bytes, BW 100000 Kbit/sec, DLY 100 usec,
      reliability 255/255, txload 1/255, rxload 1/255
```

p. From the R1 command prompt, ping PC-A again. You should be successful.

Reflection

1. Why would you enable port security on a switch?

2. Why should unused ports on a switch be disabled?

Router Interface Summary Table

Router Interface Summary				
Router Model	**Ethernet Interface #1**	**Ethernet Interface #2**	**Serial Interface #1**	**Serial Interface #2**
1800	Fast Ethernet 0/0 (F0/0)	Fast Ethernet 0/1 (F0/1)	Serial 0/0/0 (S0/0/0)	Serial 0/0/1 (S0/0/1)
1900	Gigabit Ethernet 0/0 (G0/0)	Gigabit Ethernet 0/1 (G0/1)	Serial 0/0/0 (S0/0/0)	Serial 0/0/1 (S0/0/1)
2801	Fast Ethernet 0/0 (F0/0)	Fast Ethernet 0/1 (F0/1)	Serial 0/1/0 (S0/1/0)	Serial 0/1/1 (S0/1/1)
2811	Fast Ethernet 0/0 (F0/0)	Fast Ethernet 0/1 (F0/1)	Serial 0/0/0 (S0/0/0)	Serial 0/0/1 (S0/0/1)
2900	Gigabit Ethernet 0/0 (G0/0)	Gigabit Ethernet 0/1 (G0/1)	Serial 0/0/0 (S0/0/0)	Serial 0/0/1 (S0/0/1)
Note: To find out how the router is configured, look at the interfaces to identify the type of router and how many interfaces the router has. There is no way to effectively list all the combinations of configurations for each router class. This table includes identifiers for the possible combinations of Ethernet and Serial interfaces in the device. The table does not include any other type of interface, even though a specific router may contain one. An example of this might be an ISDN BRI interface. The string in parenthesis is the legal abbreviation that can be used in Cisco IOS commands to represent the interface.				

2.3.1.1 Class Activity – Switch Trio

Objective

Verify the Layer 2 configuration of a switch port connected to an end station.

Scenario

You are the network administrator for a small- to medium-sized business. Corporate headquarters for your business has mandated that on all switches in all offices, security must be implemented. The memorandum delivered to you this morning states:

> "By Monday, April 18, 20xx, the first three ports of all configurable switches located in all offices must be secured with MAC addresses — one address will be reserved for the PC, one address will be reserved for the laptop in the office, and one address will be reserved for the office server.
>
> If a port's security is breached, we ask you to shut it down until the reason for the breach can be certified.
>
> Please implement this policy no later than the date stated in this memorandum. For questions, call 1.800.555.1212. Thank you. The Network Management Team"

Work with a partner in the class and create a Packet Tracer example to test this new security policy. Once you have created your file, test it with, at least, one device to ensure it is operational or validated.

Save your work and be prepared to share it with the entire class.

Reflection

1. Why would one port on a switch be secured on a switch using these scenario parameters (and not all the ports on the same switch)?

   _____

2. Why would a network administrator use a network simulator to create, configure, and validate a security plan, instead of using the small- to medium-sized business' actual, physical equipment?

   _____

Chapter 3 — VLANs

3.0.1.2 Class Activity – Vacation Station

Objective

Explain the purpose of VLANs in a switched network.

Scenario

You have purchased a vacation home at the beach for rental purposes. There are three, identical floors on each level of the home. Each floor offers one digital television for renters to use.

According to the local Internet service provider, only three stations may be offered within a television package. It is your job to decide which television packages you offer your guests.

- Divide the class into groups of three students per group.
- Choose three different stations to make one subscription package for each floor of your rental home.
- Complete the table for this activity.

Share your completed group-reflection answers with the class.

Television Station Subscription Package – Floor 1		
Local News	**Sports**	**Weather**
Home Improvement	**Movies**	**History**
Television Station Subscription Package – Floor 2		
Local News	**Sports**	**Weather**
Home Improvement	**Movies**	**History**
Television Station Subscription Package – Floor 3		
Local News	**Sports**	**Weather**
Home Improvement	**Movies**	**History**

Reflection

1. What were some of the criteria you used to select the final three stations?

_____

2. Why do you think this Internet service provider offers different television station options to subscribers? Why not offer all stations to all subscribers?

_____

3. Compare this scenario to data communications and networks for small- to medium-sized businesses. Why would it be a good idea to divide your small- to medium-sized business networks into logical and physical groups?

_____

3.2.2.5 Lab - Configuring VLANs and Trunking

Topology

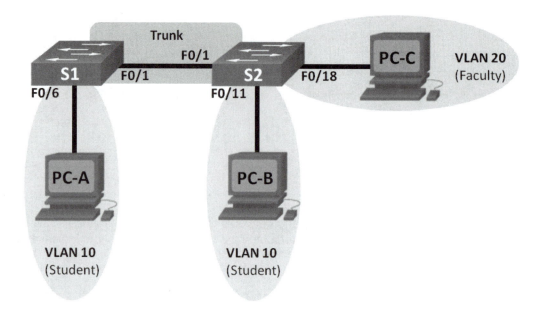

Addressing Table

Device	Interface	IP Address	Subnet Mask	Default Gateway
S1	VLAN 1	192.168.1.11	255.255.255.0	N/A
S2	VLAN 1	192.168.1.12	255.255.255.0	N/A
PC-A	NIC	192.168.10.3	255.255.255.0	192.168.10.1
PC-B	NIC	192.168.10.4	255.255.255.0	192.168.10.1
PC-C	NIC	192.168.20.3	255.255.255.0	192.168.20.1

Objectives

Part 1: Build the Network and Configure Basic Device Settings

Part 2: Create VLANs and Assign Switch Ports

Part 3: Maintain VLAN Port Assignments and the VLAN Database

Part 4: Configure an 802.1Q Trunk between the Switches

Part 5: Delete the VLAN Database

Background / Scenario

Modern switches use virtual local-area networks (VLANs) to improve network performance by separating large Layer 2 broadcast domains into smaller ones. VLANs can also be used as a security measure by controlling which hosts can communicate. In general, VLANs make it easier to design a network to support the goals of an organization.

VLAN trunks are used to span VLANs across multiple devices. Trunks allow the traffic from multiple VLANS to travel over a single link, while keeping the VLAN identification and segmentation intact.

In this lab, you will create VLANs on both switches in the topology, assign VLANs to switch access ports, verify that VLANs are working as expected, and then create a VLAN trunk between the two switches to allow hosts in the same VLAN to communicate through the trunk, regardless of which switch the host is actually attached to.

Note: The switches used are Cisco Catalyst 2960s with Cisco IOS Release 15.0(2) (lanbasek9 image). Other switches and Cisco IOS versions can be used. Depending on the model and Cisco IOS version, the commands available and output produced might vary from what is shown in the labs.

Note: Ensure that the switches have been erased and have no startup configurations. If you are unsure contact your instructor.

Required Resources

- 2 Switches (Cisco 2960 with Cisco IOS Release 15.0(2) lanbasek9 image or comparable)
- 3 PCs (Windows 7, Vista, or XP with terminal emulation program, such as Tera Term)
- Console cables to configure the Cisco IOS devices via the console ports
- Ethernet cables as shown in the topology

Part 1:   Build the Network and Configure Basic Device Settings

In Part 1, you will set up the network topology and configure basic settings on the PC hosts and switches.

Step 1:   Cable the network as shown in the topology.

Attach the devices as shown in the topology diagram, and cable as necessary.

Step 2:   Initialize and reload the switches as necessary.

Step 3:   Configure basic settings for each switch.

a.   Disable DNS lookup.

b.   Configure device name as shown in the topology.

c.   Assign **class** as the privileged EXEC password.

d.   Assign **cisco** as the console and vty passwords and enable login for console and vty lines.

e.   Configure **logging synchronous** for the console line.

f.   Configure a MOTD banner to warn users that unauthorized access is prohibited.

g.   Configure the IP address listed in the Addressing Table for VLAN 1 on both switches.

h.   Administratively deactivate all unused ports on the switch.

i.   Copy the running configuration to the startup configuration.

Step 4:   **Configure PC hosts.**

Refer to the Addressing Table for PC host address information.

Step 5:   **Test connectivity.**

Verify that the PC hosts can ping one another.

Note: It may be necessary to disable the PCs firewall to ping between PCs.

Can PC-A ping PC-B?   _____

Can PC-A ping PC-C?   _____

Can PC-A ping S1?   _____

Can PC-B ping PC-C?   _____

Can PC-B ping S2?   _____

Can PC-C ping S2?   _____

Can S1 ping S2?   _____

If you answered no to any of the above questions, why were the pings unsuccessful?

_____

_____

Part 2:   Create VLANs and Assign Switch Ports

In Part 2, you will create student, faculty, and management VLANs on both switches. You will then assign the VLANs to the appropriate interface. The **show vlan** command is used to verify your configuration settings.

Step 1:   **Create VLANs on the switches.**

a.   Create the VLANs on S1.

```
S1(config)# vlan 10
S1(config-vlan)# name Student
S1(config-vlan)# vlan 20
S1(config-vlan)# name Faculty
S1(config-vlan)# vlan 99
S1(config-vlan)# name Management
S1(config-vlan)# end
```

b.   Create the same VLANs on S2.

c.   Issue the **show vlan** command to view the list of VLANs on S1.

```
S1# show vlan
```

VLAN	Name	Status	Ports
1	default	active	Fa0/1, Fa0/2, Fa0/3, Fa0/4
			Fa0/5, Fa0/6, Fa0/7, Fa0/8
			Fa0/9, Fa0/10, Fa0/11, Fa0/12
			Fa0/13, Fa0/14, Fa0/15, Fa0/16
			Fa0/17, Fa0/18, Fa0/19, Fa0/20
			Fa0/21, Fa0/22, Fa0/23, Fa0/24
			Gi0/1, Gi0/2
10	Student	active	
20	Faculty	active	
99	Management	active	
1002	fddi-default	act/unsup	
1003	token-ring-default	act/unsup	
1004	fddinet-default	act/unsup	
1005	trnet-default	act/unsup	

VLAN	Type	SAID	MTU	Parent	RingNo	BridgeNo	Stp	BrdgMode	Trans1	Trans2
1	enet	100001	1500	-	-	-	-	-	0	0
10	enet	100010	1500	-	-	-	-	-	0	0
20	enet	100020	1500	-	-	-	-	-	0	0
99	enet	100099	1500	-	-	-	-	-	0	0

VLAN	Type	SAID	MTU	Parent	RingNo	BridgeNo	Stp	BrdgMode	Trans1	Trans2
1002	fddi	101002	1500	-	-	-	-	-	0	0
1003	tr	101003	1500	-	-	-	-	-	0	0
1004	fdnet	101004	1500	-	-	-	ieee	-	0	0
1005	trnet	101005	1500	-	-	-	ibm	-	0	0

```
Remote SPAN VLANs
-------------------------------------------------------------------------


Primary Secondary Type              Ports
------- --------- ----------------- ----------------------------------------
```

What is the default VLAN? _____

What ports are assigned to the default VLAN?

_____

Step 2:   Assign VLANs to the correct switch interfaces.

a.   Assign VLANs to the interfaces on S1.

   1)   Assign PC-A to the Student VLAN.

```
S1(config)# interface f0/6
S1(config-if)# switchport mode access
S1(config-if)# switchport access vlan 10
```

   2)   Move the switch IP address VLAN 99.

```
S1(config)# interface vlan 1
S1(config-if)# no ip address
S1(config-if)# interface vlan 99
S1(config-if)# ip address 192.168.1.11 255.255.255.0
S1(config-if)# end
```

b.   Issue the **show vlan brief** command and verify that the VLANs are assigned to the correct interfaces.

```
S1# show vlan brief
```

```
VLAN Name                             Status     Ports
---- -------------------------------- ---------- --------------------------------
1    default                          active     Fa0/1, Fa0/2, Fa0/3, Fa0/4
                                                 Fa0/5, Fa0/7, Fa0/8, Fa0/9
                                                 Fa0/10, Fa0/11, Fa0/12, Fa0/13
                                                 Fa0/14, Fa0/15, Fa0/16, Fa0/17
                                                 Fa0/18, Fa0/19, Fa0/20, Fa0/21
                                                 Fa0/22, Fa0/23, Fa0/24, Gi0/1
                                                 Gi0/2
10   Student                          active     Fa0/6
20   Faculty                          active
99   Management                       active
1002 fddi-default                     act/unsup
1003 token-ring-default               act/unsup
1004 fddinet-default                  act/unsup
1005 trnet-default                    act/unsup
```

c.   Issue the **show ip interfaces brief** command.

What is the status of VLAN 99? Why?

_____

d. Use the Topology to assign VLANs to the appropriate ports on S2.

e. Remove the IP address for VLAN 1 on S2.

f. Configure an IP address for VLAN 99 on S2 according to the Addressing Table.

g. Use the **show vlan brief** command to verify that the VLANs are assigned to the correct interfaces.

```
S2# show vlan brief


VLAN Name                             Status    Ports
---- -------------------------------- --------- -------------------------------
1    default                          active    Fa0/1, Fa0/2, Fa0/3, Fa0/4
                                                Fa0/5, Fa0/6, Fa0/7, Fa0/8
                                                Fa0/9, Fa0/10, Fa0/12, Fa0/13
                                                Fa0/14, Fa0/15, Fa0/16, Fa0/17
                                                Fa0/19, Fa0/20, Fa0/21, Fa0/22
                                                Fa0/23, Fa0/24, Gi0/1, Gi0/2
10   Student                          active    Fa0/11
20   Faculty                          active    Fa0/18
99   Management                       active
1002 fddi-default                     act/unsup
1003 token-ring-default               act/unsup
1004 fddinet-default                  act/unsup
1005 trnet-default                    act/unsup
```

Is PC-A able to ping PC-B? Why?

_____

Is S1 able to ping S2? Why?

_____

Part 3:   Maintain VLAN Port Assignments and the VLAN Database

In Part 3, you will change VLAN assignments to ports and remove VLANs from the VLAN database.

Step 1:   Assign a VLAN to multiple interfaces.

a.   On S1, assign interfaces F0/11 – 24 to VLAN 10.

```
S1(config)# interface range f0/11-24
S1(config-if-range)# switchport mode access
S1(config-if-range)# switchport access vlan 10
S1(config-if-range)# end
```

b.   Issue the **show vlan brief** command to verify VLAN assignments.

c.   Reassign F0/11 and F0/21 to VLAN 20.

d.   Verify that VLAN assignments are correct.

Step 2:   Remove a VLAN assignment from an interface.

a. Use the **no switchport access vlan** command to remove the VLAN 10 assignment to F0/24.

```
S1(config)# interface f0/24
S1(config-if)# no switchport access vlan
S1(config-if)# end
```

b. Verify that the VLAN change was made.

Which VLAN is F0/24 is now associated with?

_____

Step 3:   Remove a VLAN ID from the VLAN database.

a. Add VLAN 30 to interface F0/24 without issuing the VLAN command.

```
S1(config)# interface f0/24
S1(config-if)# switchport access vlan 30
% Access VLAN does not exist. Creating vlan 30
```

Note: Current switch technology no longer requires that the **vlan** command be issued to add a VLAN to the database. By assigning an unknown VLAN to a port, the VLAN adds to the VLAN database.

b. Verify that the new VLAN is displayed in the VLAN table.

```
S1# show vlan brief
```

```
VLAN Name                             Status    Ports
---- -------------------------------- --------- -------------------------------
1    default                          active    Fa0/1, Fa0/2, Fa0/3, Fa0/4
                                                Fa0/5, Fa0/6, Fa0/7, Fa0/8
                                                Fa0/9, Fa0/10, Gi0/1, Gi0/2
10   Student                          active    Fa0/12, Fa0/13, Fa0/14, Fa0/15
                                                Fa0/16, Fa0/17, Fa0/18, Fa0/19
                                                Fa0/20, Fa0/22, Fa0/23
20   Faculty                          active    Fa0/11, Fa0/21
30   VLAN0030                         active    Fa0/24
99   Management                       active
1002 fddi-default                     act/unsup
1003 token-ring-default               act/unsup
1004 fddinet-default                  act/unsup
1005 trnet-default                    act/unsup
```

What is the default name of VLAN 30?

_____

c.   Use the **no vlan 30** command to remove VLAN 30 from the VLAN database.

   S1(config)# **no vlan 30**

   S1(config)# **end**

d.   Issue the **show vlan brief** command. F0/24 was assigned to VLAN 30.

   After deleting VLAN 30, what VLAN is port F0/24 assigned to? What happens to the traffic destined to the host attached to F0/24?

_____

   S1# **show vlan brief**

```
VLAN Name                            Status    Ports
---- -------------------------------- --------- -------------------------------
1    default                          active    Fa0/1, Fa0/2, Fa0/3, Fa0/4
                                                Fa0/5, Fa0/6, Fa0/7, Fa0/8
                                                Fa0/9, Fa0/10, Gi0/1, Gi0/2
10   Student                          active    Fa0/12, Fa0/13, Fa0/14, Fa0/15
                                                Fa0/16, Fa0/17, Fa0/18, Fa0/19
                                                Fa0/20, Fa0/22, Fa0/23
20   Faculty                          active    Fa0/11, Fa0/21
99   Management                       active
1002 fddi-default                     act/unsup
1003 token-ring-default               act/unsup
1004 fddinet-default                  act/unsup
1005 trnet-default                    act/unsup
```

e.   Issue the **no switchport access vlan** command on interface F0/24.

_____

f.   Issue the **show vlan brief** command to determine the VLAN assignment for F0/24. To which VLAN is F0/24 assigned?

_____

Note: Before removing a VLAN from the database, it is recommended that you reassign all the ports assigned to that VLAN.

Why should you reassign a port to another VLAN before removing the VLAN from the VLAN database?

_____

_____

_____

Part 4: Configure an 802.1Q Trunk Between the Switches

In Part 4, you will configure interface F0/1 to use the Dynamic Trunking Protocol (DTP) to allow it to negotiate the trunk mode. After this has been accomplished and verified, you will disable DTP on interface F0/1 and manually configure it as a trunk.

Step 1: Use DTP to initiate trunking on F0/1.

The default DTP mode of a 2960 switch port is dynamic auto. This allows the interface to convert the link to a trunk if the neighboring interface is set to trunk or dynamic desirable mode.

a. Set F0/1 on S1 to negotiate trunk mode.

```
S1(config)# interface f0/1

S1(config-if)# switchport mode dynamic desirable

*Mar  1 05:07:28.746: %LINEPROTO-5-UPDOWN: Line protocol on Interface Vlan1, changed
state to down

*Mar  1 05:07:29.744: %LINEPROTO-5-UPDOWN: Line protocol on Interface FastEthernet0/1,
changed state to down

S1(config-if)#

*Mar  1 05:07:32.772: %LINEPROTO-5-UPDOWN: Line protocol on Interface FastEthernet0/1,
changed state to up

S1(config-if)#

*Mar  1 05:08:01.789: %LINEPROTO-5-UPDOWN: Line protocol on Interface Vlan99, changed
state to up

*Mar  1 05:08:01.797: %LINEPROTO-5-UPDOWN: Line protocol on Interface Vlan1, changed
state to up
```

You should also receive link status messages on S2.

```
S2#

*Mar  1 05:07:29.794: %LINEPROTO-5-UPDOWN: Line protocol on Interface FastEthernet0/1,
changed state to down

S2#

*Mar  1 05:07:32.823: %LINEPROTO-5-UPDOWN: Line protocol on Interface FastEthernet0/1,
changed state to up

S2#

*Mar  1 05:08:01.839: %LINEPROTO-5-UPDOWN: Line protocol on Interface Vlan99, changed
state to up

*Mar  1 05:08:01.850: %LINEPROTO-5-UPDOWN: Line protocol on Interface Vlan1, changed
state to up
```

b. Issue the **show vlan brief** command on S1 and S2. Interface F0/1 is no longer assigned to VLAN 1.
    Trunked interfaces are not listed in the VLAN table.

```
S1# show vlan brief
```

VLAN	Name	Status	Ports
1	default	active	Fa0/2, Fa0/3, Fa0/4, Fa0/5
			Fa0/7, Fa0/8, Fa0/9, Fa0/10
			Fa0/24, Gi0/1, Gi0/2
10	Student	active	Fa0/6, Fa0/12, Fa0/13, Fa0/14
			Fa0/15, Fa0/16, Fa0/17, Fa0/18
			Fa0/19, Fa0/20, Fa0/22, Fa0/23
20	Faculty	active	Fa0/11, Fa0/21
99	Management	active	
1002	fddi-default	act/unsup	
1003	token-ring-default	act/unsup	
1004	fddinet-default	act/unsup	
1005	trnet-default	act/unsup	

c. Issue the **show interfaces trunk** command to view trunked interfaces. Notice that the mode on S1 is set
    to desirable, and the mode on S2 is set to auto.

```
S1# show interfaces trunk
```

Port	Mode	Encapsulation	Status	Native vlan
Fa0/1	desirable	802.1q	trunking	1

Port	Vlans allowed on trunk
Fa0/1	1-4094

Port	Vlans allowed and active in management domain
Fa0/1	1,10,20,99

Port	Vlans in spanning tree forwarding state and not pruned
Fa0/1	1,10,20,99

```
S2# show interfaces trunk
```

Port	Mode	Encapsulation	Status	Native vlan
Fa0/1	auto	802.1q	trunking	1

Port	Vlans allowed on trunk
Fa0/1	1-4094

```
Port          Vlans allowed and active in management domain
Fa0/1         1,10,20,99


Port          Vlans in spanning tree forwarding state and not pruned
Fa0/1         1,10,20,99
```

Note: By default, all VLANs are allowed on a trunk. The **switchport trunk** command allows you to control what VLANs have access to the trunk. For this lab, keep the default settings which allows all VLANs to traverse F0/1.

d. Verify that VLAN traffic is traveling over trunk interface F0/1.

Can S1 ping S2? _____

Can PC-A ping PC-B? _____

Can PC-A ping PC-C? _____

Can PC-B ping PC-C? _____

Can PC-A ping S1? _____

Can PC-B ping S2? _____

Can PC-C ping S2? _____

If you answered no to any of the above questions, explain below.

_____

_____

Step 2: Manually configure trunk interface F0/1.

The **switchport mode trunk** command is used to manually configure a port as a trunk. This command should be issued on both ends of the link.

a. Change the switchport mode on interface F0/1 to force trunking. Make sure to do this on both switches.

```
S1(config)# interface f0/1
S1(config-if)# switchport mode trunk
```

b. Issue the **show interfaces trunk** command to view the trunk mode. Notice that the mode changed from **desirable** to **on**.

```
S2# show interfaces trunk
```

```
Port          Mode         Encapsulation   Status      Native vlan
Fa0/1         on           802.1q          trunking    99
```

```
Port          Vlans allowed on trunk
Fa0/1         1-4094


Port          Vlans allowed and active in management domain
Fa0/1         1,10,20,99


Port          Vlans in spanning tree forwarding state and not pruned
Fa0/1         1,10,20,99
```

Why might you want to manually configure an interface to trunk mode instead of using DTP?

_____

_____

Part 5:   **Delete the VLAN Database**

In Part 5, you will delete the VLAN Database from the switch. It is necessary to do this when initializing a switch back to its default settings.

Step 1:   **Determine if the VLAN database exists.**

Issue the **show flash** command to determine if a **vlan.dat** file exists in flash.

```
S1# show flash


Directory of flash:/


    2  -rwx         1285    Mar 1 1993 00:01:24 +00:00  config.text
    3  -rwx        43032    Mar 1 1993 00:01:24 +00:00  multiple-fs
    4  -rwx            5    Mar 1 1993 00:01:24 +00:00  private-config.text
    5  -rwx     11607161    Mar 1 1993 02:37:06 +00:00  c2960-lanbasek9-mz.150-2.SE.bin
    6  -rwx          736    Mar 1 1993 00:19:41 +00:00  vlan.dat


32514048 bytes total (20858880 bytes free)
```

Note: If there is a **vlan.dat** file located in flash, then the VLAN database does not contain its default settings.

Step 2:   **Delete the VLAN database.**

a. Issue the **delete vlan.dat** command to delete the vlan.dat file from flash and reset the VLAN database back to its default settings. You will be prompted twice to confirm that you want to delete the vlan.dat file. Press Enter both times.

```
S1# delete vlan.dat

Delete filename [vlan.dat]?

Delete flash:/vlan.dat? [confirm]
```

```
S1#
```

b. Issue the **show flash** command to verify that the vlan.dat file has been deleted.

```
S1# show flash

Directory of flash:/

    2  -rwx        1285    Mar 1 1993 00:01:24 +00:00  config.text
    3  -rwx       43032    Mar 1 1993 00:01:24 +00:00  multiple-fs
    4  -rwx           5    Mar 1 1993 00:01:24 +00:00  private-config.text
    5  -rwx    11607161    Mar 1 1993 02:37:06 +00:00  c2960-lanbasek9-mz.150-2.SE.bin

32514048 bytes total (20859904 bytes free)
```

To initialize a switch back to its default settings, what other commands are needed?

_____

_____

Reflection

1. What is needed to allow hosts on VLAN 10 to communicate to hosts on VLAN 20?

_____

_____

2. What are some primary benefits that an organization can receive through effective use of VLANs?

_____

_____

_____

_____

3.2.4.9 Lab - Troubleshooting VLAN Configurations

Topology

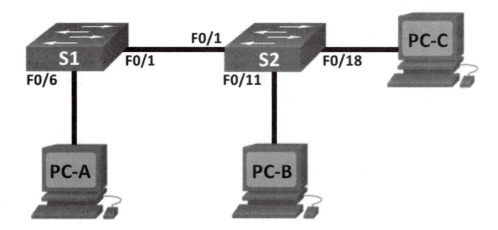

Addressing Table

Device	Interface	IP Address	Subnet Mask	Default Gateway
S1	VLAN 1	192.168.1.2	255.255.255.0	N/A
S2	VLAN 1	192.168.1.3	255.255.255.0	N/A
PC-A	NIC	192.168.10.2	255.255.255.0	192.168.10.1
PC-B	NIC	192.168.10.3	255.255.255.0	192.168.10.1
PC-C	NIC	192.168.20.3	255.255.255.0	192.168.20.1

Switch Port Assignment Specifications

Ports	Assignment	Network
F0/1	802.1Q Trunk	N/A
F0/6-12	VLAN 10 – Students	192.168.10.0/24
F0/13-18	VLAN 20 – Faculty	192.168.20.0/24
F0/19-24	VLAN 30 – Guest	192.168.30.0/24

Objectives

Part 1: Build the Network and Configure Basic Device Settings

Part 2: Troubleshoot VLAN 10

Part 3: Troubleshoot VLAN 20

Background / Scenario

VLANs provide logical segmentation within an internetwork and improve network performance by separating large broadcast domains into smaller ones. By separating hosts into different networks, VLANs can be used to control which hosts can communicate. In this lab, a school has decided to implement VLANs in order to separate traffic from different end users. The school is using 802.1Q trunking to facilitate VLAN communication between switches.

The S1 and S2 switches have been configured with VLAN and trunking information. Several errors in the configuration have resulted in connectivity issues. You have been asked to troubleshoot and correct the configuration errors and document your work.

Note: The switches used with this lab are Cisco Catalyst 2960s with Cisco IOS Release 15.0(2) (lanbasek9 image). Other switches and Cisco IOS versions can be used. Depending on the model and Cisco IOS version, the commands available and output produced might vary from what is shown in the labs.

Note: Make sure that the switches have been erased and have no startup configurations. If you are unsure, contact your instructor.

Required Resources

- 2 Switches (Cisco 2960 with Cisco IOS Release 15.0(2) lanbasek9 image or comparable)
- 3 PCs (Windows 7, Vista, or XP with terminal emulation program, such as Tera Term)
- Console cables to configure the Cisco IOS devices via the console ports
- Ethernet cables as shown in the topology

Part 1: Build the Network and Configure Basic Device Settings

In Part 1, you will set up the network topology and configure the switches with some basic settings, such as passwords and IP addresses. Preset VLAN-related configurations, which contain errors, are provided for you for the initial switch configurations. You will also configure the IP settings for the PCs in the topology.

Step 1: **Cable the network as shown in the topology.**

Step 2: **Configure PC hosts.**

Step 3: **Initialize and reload the switches as necessary.**

Step 4: **Configure basic settings for each switch.**

a. Disable DNS lookup.

b. Configure IP address in Addressing Table.

c. Assign **cisco** as the console and vty passwords and enable login for console and vty lines.

d. Assign **class** as the privileged EXEC password.

e. Configure **logging synchronous** to prevent console messages from interrupting command entry.

Step 5:   **Load switch configurations.**

The configurations for the switches S1 and S2 are provided for you. There are errors within these configurations, and it is your job to determine the incorrect configurations and correct them.

Switch S1 Configuration:

```
hostname S1
vlan 10
name Students
vlan 2
!vlan 20
name Faculty
vlan 30
name Guest
interface range f0/1-24
switchport mode access
shutdown




interface range f0/7-12


switchport access vlan 10
interface range f0/13-18
switchport access vlan 2


interface range f0/19-24
switchport access vlan 30
end
```

Switch S2 Configuration:

```
hostname S2
vlan 10
Name Students
vlan 20
Name Faculty
vlan 30
Name Guest
interface f0/1
switchport mode trunk
switchport trunk allowed vlan 1,10,2,30
```

```
interface range f0/2-24
switchport mode access
shutdown


interface range f0/13-18
switchport access vlan 20
interface range f0/19-24
switchport access vlan 30
shutdown
end
```

Step 6:   **Copy the running configuration to the startup configuration.**

Part 2:   Troubleshoot VLAN 10

In Part 2, you must examine VLAN 10 on S1 and S2 to determine if it is configured correctly. You will trouble-shoot the scenario until connectivity is established.

Step 1:   **Troubleshoot VLAN 10 on S1.**

a. Can PC-A ping PC-B? _____

b. After verifying that PC-A was configured correctly, examine the S1 switch to find possible configuration errors by viewing a summary of the VLAN information. Enter the **show vlan brief** command.

c. Are there any problems with the VLAN configuration?

_____

d. Examine the switch for trunk configurations using the **show interfaces trunk** and the **show interface f0/1 switchport** commands.

e. Are there any problems with the trunking configuration?

_____

f. Examine the running configuration of the switch to find possible configuration errors.

Are there any problems with the current configuration?

_____

g. Correct the errors found regarding F0/1 and VLAN 10 on S1. Record the commands used in the space below.

_____

_____

_____

_____

h. Verify the commands had the desired effects by issuing the appropriate **show** commands.

i. Can PC-A ping PC-B? _____

Step 2: **Troubleshoot VLAN 10 on S2.**

a. Using the previous commands, examine the S2 switch to find possible configuration errors.

Are there any problems with the current configuration?

_____

b. Correct the errors found regarding interfaces and VLAN 10 on S2. Record the commands below.

_____

_____

_____

_____

_____

_____

c. Can PC-A ping PC-B? _____

Part 3: **Troubleshoot VLAN 20**

In Part 3, you must examine VLAN 20 on S1 and S2 to determine if it is configured correctly. To verify functionality, you will reassign PC-A into VLAN 20, and then troubleshoot the scenario until connectivity is established.

Step 1: **Assign PC-A to VLAN 20.**

a. On PC-A, change the IP address to 192.168.20.2/24 with a default gateway of 192.168.20.1.

b. On S1, assign the port for PC-A to VLAN 20. Write the commands needed to complete the configuration.

_____

_____

c. Verify that the port for PC-A has been assigned to VLAN 20.

d. Can PC-A ping PC-C? _____

Step 2:   **Troubleshoot VLAN 20 on S1.**

a.   Using the previous commands, examine the S1 switch to find possible configuration errors.

Are there any problems with the current configuration?

_____

_____

b.   Correct the errors found regarding VLAN 20.

c.   Can PC-A ping PC-C? _____

Step 3:   **Troubleshoot VLAN 20 on S2.**

a.   Using the previous commands, examine the S2 switch to find possible configuration errors.

Are there any problems with the current configuration?

_____

b. Correct the errors found regarding VLAN 20. Record the commands used below.

_____

_____

_____

_____

_____

c. Can PC-A ping PC-C? _____

Note: It may be necessary to disable the PC firewall to ping between PCs.

Reflection

1. Why is a correctly configured trunk port critical in a multi-VLAN environment?

_____

_____

2. Why would a network administrator limit traffic for specific VLANs on a trunk port?

_____

_____

3.3.2.2 Lab – Implementing VLAN Security

Topology

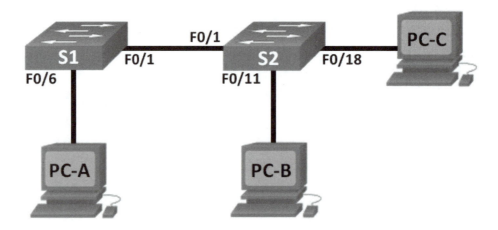

Addressing Table

Device	Interface	IP Address	Subnet Mask	Default Gateway
S1	VLAN 99	172.17.99.11	255.255.255.0	172.17.99.1
S2	VLAN 99	172.17.99.12	255.255.255.0	172.17.99.1
PC-A	NIC	172.17.99.3	255.255.255.0	172.17.99.1
PC-B	NIC	172.17.10.3	255.255.255.0	172.17.10.1
PC-C	NIC	172.17.99.4	255.255.255.0	172.17.99.1

VLAN Assignments

VLAN	Name
10	Data
99	Management&Native
999	BlackHole

Objectives

Part 1: Build the Network and Configure Basic Device Settings

Part 2: Implement VLAN Security on the Switches

Background / Scenario

Best practice dictates configuring some basic security settings for both access and trunk ports on switches. This will help guard against VLAN attacks and possible sniffing of network traffic within the network.

In this lab, you will configure the network devices in the topology with some basic settings, verify connectivity and then apply more stringent security measures on the switches. You will examine how Cisco switches behave by using various **show** commands. You will then apply security measures.

Note: The switches used with this lab are Cisco Catalyst 2960s with Cisco IOS Release 15.0(2) (lanbasek9 image). Other switches and Cisco IOS versions can be used. Depending on the model and Cisco IOS version, the commands available and output produced might vary from what is shown in the labs.

Note: Make sure that the switches have been erased and have no startup configurations. If you are unsure, contact your instructor.

Required Resources

- 2 Switches (Cisco 2960 with Cisco IOS Release 15.0(2) lanbasek9 image or comparable)
- 3 PCs (Windows 7, Vista, or XP with terminal emulation program, such as Tera Term)
- Console cables to configure the Cisco IOS devices via the console ports
- Ethernet cables as shown in the topology

Part 1: Build the Network and Configure Basic Device Settings

In Part 1, you will configure basic settings on the switches and PCs. Refer to the Addressing Table for device names and address information.

Step 1: Cable the network as shown in the topology.

Step 2: Initialize and reload the switches.

Step 3: Configure IP addresses on PC-A, PC-B, and PC-C.

Refer to the Addressing Table for PC address information.

Step 4: Configure basic settings for each switch.

a. Disable DNS lookup.

b. Configure the device names as shown in the topology.

c. Assign **class** as the privileged EXEC mode password.

d. Assign **cisco** as the console and VTY password and enable login for console and vty lines.

e. Configure synchronous logging for console and vty lines.

Step 5: Configure VLANs on each switch.

a. Create and name VLANs according to the VLAN Assignments table.

b. Configure the IP address listed in the Addressing Table for VLAN 99 on both switches.

c. Configure F0/6 on S1 as an access port and assign it to VLAN 99.

    d. Configure F0/11 on S2 as an access port and assign it to VLAN 10.

    e. Configure F0/18 on S2 as an access port and assign it to VLAN 99.

    f. Issue **show vlan brief** command to verify VLAN and port assignments.

To which VLAN would an unassigned port, such as F0/8 on S2, belong?

_____

Step 6: Configure basic switch security.

    a. Configure a MOTD banner to warn users that unauthorized access is prohibited.

    b. Encrypt all passwords.

    c. Shut down all unused physical ports.

    d. Disable the basic web service running.

```
S1(config)# no ip http server
S2(config)# no ip http server
```

    e. Copy the running configuration to startup configuration.

Step 7:   **Verify connectivity between devices and VLAN information.**

a. From a command prompt on PC-A, ping the management address of S1. Were the pings successful? Why?

_____

_____

b. From S1, ping the management address of S2. Were the pings successful? Why?

_____

_____

c. From a command prompt on PC-B, ping the management addresses on S1 and S2 and the IP address of PC-A and PC-C. Were your pings successful? Why?

_____

_____

d. From a command prompt on PC-C, ping the management addresses on S1 and S2. Were you successful? Why?

_____

_____

Note: It may be necessary to disable the PC firewall to ping between PCs.

Part 2:   **Implement VLAN Security on the Switches**

Step 1:   **Configure trunk ports on S1 and S2.**

a. Configure port F0/1 on S1 as a trunk port.

```
S1(config)# interface f0/1
S1(config-if)# switchport mode trunk
```

b. Configure port F0/1 on S2 as a trunk port.

```
S2(config)# interface f0/1
S2(config-if)# switchport mode trunk
```

c.   Verify trunking on S1 and S2. Issue the **show interface trunk** command on both switches.

```
S1# show interface trunk


Port          Mode               Encapsulation  Status        Native vlan
Fa0/1         on                 802.1q         trunking      1


Port          Vlans allowed on trunk
Fa0/1         1-4094


Port          Vlans allowed and active in management domain
Fa0/1         1,10,99,999


Port          Vlans in spanning tree forwarding state and not pruned
Fa0/1         1,10,99,999
```

Step 2:   **Change the native VLAN for the trunk ports on S1 and S2.**

Changing the native VLAN for trunk ports from VLAN 1 to another VLAN is a good practice for security.

a.   What is the current native VLAN for the S1 and S2 F0/1 interfaces?

_____

b.   Configure the native VLAN on the S1 F0/1 trunk interface to Management&Native VLAN 99.

```
S1# config t
S1(config)# interface f0/1
S1(config-if)# switchport trunk native vlan 99
```

c.   Wait a few seconds. You should start receiving error messages on the console session of S1. What does the %CDP-4-NATIVE_VLAN_MISMATCH: message mean?

_____

d.   Configure the native VLAN on the S2 F0/1 trunk interface to VLAN 99.

```
S2(config)# interface f0/1
S2(config-if)# switchport trunk native vlan 99
```

e. Verify that the native VLAN is now 99 on both switches. S1 output is shown below.

```
S1# show interface trunk

Port          Mode              Encapsulation  Status        Native vlan
Fa0/1         on                802.1q         trunking      99


Port          Vlans allowed on trunk
Fa0/1         1-4094


Port          Vlans allowed and active in management domain
Fa0/1         1,10,99,999


Port           Vlans in spanning tree forwarding state and not pruned
Fa0/1         10,999
```

Step 3: Verify that traffic can successfully cross the trunk link.

a. From a command prompt on PC-A, ping the management address of S1. Were the pings successful? Why?

_____

_____

b. From the console session on S1, ping the management address of S2. Were the pings successful? Why?

_____

c. From a command prompt on PC-B, ping the management addresses on S1 and S2 and the IP address of PC-A and PC-C. Were your pings successful? Why?

_____

d. From a command prompt on PC-C, ping the management addresses on S1 and S2 and the IP address of PC-A. Were you successful? Why?

_____

Step 4:   Prevent the use of DTP on S1 and S2.

Cisco uses a proprietary protocol known as the Dynamic Trunking Protocol (DTP) on its switches. Some ports automatically negotiate to trunking. A good practice is to turn off negotiation. You can see this default behavior by issuing the following command:

```
S1# show interface f0/1 switchport

Name: Fa0/1

Switchport: Enabled

Administrative Mode: trunk

Operational Mode: trunk

Administrative Trunking Encapsulation: dot1q

Operational Trunking Encapsulation: dot1q

Negotiation of Trunking: On

<Output Omitted>
```

a.   Turn off negotiation on S1.

```
S1(config)# interface f0/1

S1(config-if)# switchport nonegotiate
```

b.   Turn off negotiation on S2.

```
S2(config)# interface f0/1

S2(config-if)# switchport nonegotiate
```

c.   Verify that negotiation is off by issuing the **show interface f0/1 switchport** command on S1 and S2.

```
S1# show interface f0/1 switchport

Name: Fa0/1

Switchport: Enabled

Administrative Mode: trunk

Operational Mode: trunk

Administrative Trunking Encapsulation: dot1q

Operational Trunking Encapsulation: dot1q

Negotiation of Trunking: Off

<Output Omitted>
```

Step 5:   Secure access ports on S1 and S2.

Even though you shut down unused ports on the switches, if a device is connected to one of those ports and the interface is enabled, trunking could occur. In addition, all ports by default are in VLAN 1. A good practice is to put all unused ports in a "black hole" VLAN. In this step, you will disable trunking on all unused ports. You will also assign unused ports to VLAN 999. For the purposes of this lab, only ports 2 through 5 will be configured on both switches.

a.   Issue the **show interface f0/2 switchport** command on S1. Notice the administrative mode and state for trunking negotiation.

```
S1# show interface f0/2 switchport

Name: Fa0/2

Switchport: Enabled
```

```
Administrative Mode: dynamic auto

Operational Mode: down

Administrative Trunking Encapsulation: dot1q

Negotiation of Trunking: On

<Output Omitted>
```

b. Disable trunking on S1 access ports.

```
S1(config)# interface range f0/2 - 5
S1(config-if-range)# switchport mode access
S1(config-if-range)# switchport access vlan 999
```

c. Disable trunking on S2 access ports.

d. Verify that port F0/2 is set to access on S1.

```
S1# show interface f0/2 switchport
Name: Fa0/2

Switchport: Enabled

Administrative Mode: static access

Operational Mode: down

Administrative Trunking Encapsulation: dot1q

Negotiation of Trunking: Off

Access Mode VLAN: 999 (BlackHole)

Trunking Native Mode VLAN: 1 (default)

Administrative Native VLAN tagging: enabled

Voice VLAN: none

<Output Omitted>
```

e. Verify that VLAN port assignments on both switches are correct. S1 is shown below as an example.

```
S1# show vlan brief
```

VLAN	Name	Status	Ports
1	default	active	Fa0/7, Fa0/8, Fa0/9, Fa0/10
			Fa0/11, Fa0/12, Fa0/13, Fa0/14
			Fa0/15, Fa0/16, Fa0/17, Fa0/18
			Fa0/19, Fa0/20, Fa0/21, Fa0/22
			Fa0/23, Fa0/24, Gi0/1, Gi0/2
10	Data	active	
99	Management&Native	active	Fa0/6
999	BlackHole	active	Fa0/2, Fa0/3, Fa0/4, Fa0/5

```
1002 fddi-default                        act/unsup

1003 token-ring-default                  act/unsup

1004 fddinet-default                     act/unsup

1005 trnet-default                       act/unsup

Restrict VLANs allowed on trunk ports.
```

By default, all VLANs are allowed to be carried on trunk ports. For security reasons, it is a good practice to only allow specific desired VLANs to cross trunk links on your network.

f.   Restrict the trunk port F0/1 on S1 to only allow VLANs 10 and 99.

```
S1(config)# interface f0/1

S1(config-if)# switchport trunk allowed vlan 10,99
```

g.   Restrict the trunk port F0/1 on S2 to only allow VLANs 10 and 99.

h.   Verify the allowed VLANs. Issue a **show interface trunk** command in privileged EXEC mode on both S1 and S2.

```
S1# show interface trunk


Port        Mode              Encapsulation  Status      Native vlan
Fa0/1       on                802.1q         trunking    99


Port        Vlans allowed on trunk
Fa0/1       10,99


Port        Vlans allowed and active in management domain
Fa0/1       10,99


Port        Vlans in spanning tree forwarding state and not pruned
Fa0/1       10,99
```

What is the result?

_____

Reflection

What, if any, are the security problems with the default configuration of a Cisco switch?

_____

_____

_____

_____

3.4.1.1 Class Activity – VLAN Plan

Objective

Implement VLANs to segment a small- to medium-sized network.

Scenario

You are designing a VLAN switched network for your small- to medium- sized business.

Your business owns space on two floors of a high-rise building. The following elements need VLAN consideration and access for planning purposes:

- Management

- Finance

- Sales

- Human Resources

- Network administrator

- General visitors to your business location

You have two Cisco 3560-24PS switches.

Use a word processing software program to design your VLAN-switched network scheme.

Section 1 of your design should include the regular names of your departments, suggested VLAN names and numbers, and which switch ports would be assigned to each VLAN.

Section 2 of your design should list how security would be planned for this switched network.

Once your VLAN plan is finished, complete the reflection questions from this activity.

Save your work. Be able to explain and discuss your VLAN design with another group or with the class.

Required Resources

Word processing program

Reflection

1. What criteria did you use for assigning ports to the VLANs?

    _____

2. How could these users access your network if the switches were not physically available to general users via direct connection?

    _____

3. Could you reduce the number of switch ports assigned for general users if you used another device to connect them to the VLAN network switch? What would be affected?

    _____

Chapter 4 — Routing Concepts

4.0.1.2 Class Activity – Do We Really Need a Map?

Objectives

Describe the primary functions and features of a router.

Scenario

Using the Internet and Google Maps, located at http://maps.google.com, find a route between the capital city of your country and some other distant town, or between two places within your own city. Pay close attention to the driving or walking directions Google Maps suggests.

Notice that in many cases, Google Maps suggests more than one route between the two locations you chose. It also allows you to put additional constraints on the route, such as avoiding highways or tolls.

- Copy at least two route instructions supplied by Google Maps for this activity. Place your copies into a word processing document and save it for to use with the next step.

- Open the .pdf accompanying this modeling activity and complete it with a fellow student. Discuss the reflection questions listed on the .pdf and record your answers.

Be prepared to present your answers to the class.

Resources

- Internet connection
- Web browser
- Google Maps, http://maps.google.com/

Reflection

1. What do the individual driving, or walking based on your criteria you input, and non-highway directions look like? What exact information do they contain? How do they relate to IP routing?

2. If Google Maps offered a set of different routes, what makes this route different from the first? Why would you choose one route over another?

　　_____

3. What criteria can be used to evaluate the usefulness of a route?

　　_____

4. Is it sensible to expect that a single route can be "the best one", i.e. meeting all various requirements? Justify your answer.

　　_____

5. As a network administrator or developer, how could you use a network map, or routing table, in your daily network activities?

　　_____

4.1.1.9 Lab - Mapping the Internet

Objectives

Part 1: Determine Network Connectivity to a Destination Host

Part 2: Trace a Route to a Remote Server Using Tracert

Background / Scenario

Route tracing computer software lists the networks that data traverses from the user's originating end device to a distant destination device.

This network tool is typically executed at the command line as:

tracert `<destination network name or end device address>`

(Microsoft Windows systems)

or

traceroute `<destination network name or end device address>`

(UNIX, Linux systems, and Cisco devices, such as switches and routers)

Both **tracert** and **traceroute** determine the route taken by packets across an IP network.

The **tracert** (or **traceroute**) tool is often used for network troubleshooting. By showing a list of routers traversed, the user can identify the path taken to reach a particular destination on the network or across internetworks. Each router represents a point where one network connects to another network and through which the data packet was forwarded. The number of routers is known as the number of hops the data traveled from source to destination.

The displayed list can help identify data flow problems when trying to access a service such as a website. It can also be useful when performing tasks, such as downloading data. If there are multiple websites (mirrors) available for the same data file, one can trace each mirror to get a good idea of which mirror would be the fastest to use.

Command-line based route tracing tools are usually embedded with the operating system of the end device. This activity should be performed on a computer that has Internet access and access to a command line.

Required Resources

PC with Internet access

Part 1:   Determine Network Connectivity to a Destination Host

To trace the route to a distant network, the PC used must have a working connection to the Internet. Use the **ping** command to test whether a host is reachable. Packets of information are sent to the remote host with instructions to reply. Your local PC measures whether a response is received to each packet, and how long it takes for those packets to cross the network.

a. At the command-line prompt, type **ping www.cisco.com** to determine if it is reachable.

```
C:\>ping www.cisco.com

Pinging e144.dscb.akamaiedge.net [23.1.48.170] with 32 bytes of data:
Reply from 23.1.48.170: bytes=32 time=56ms TTL=57
Reply from 23.1.48.170: bytes=32 time=55ms TTL=57
Reply from 23.1.48.170: bytes=32 time=54ms TTL=57
Reply from 23.1.48.170: bytes=32 time=54ms TTL=57

Ping statistics for 23.1.48.170:
    Packets: Sent = 4, Received = 4, Lost = 0 (0% loss),
Approximate round trip times in milli-seconds:
    Minimum = 54ms, Maximum = 56ms, Average = 54ms
```

b. Now ping one of the Regional Internet Registry (RIR) websites located in different parts of the world to determine if it is reachable:

Africa: **www.afrinic.net**

Australia: **www.apnic.net**

South America: **www.lacnic.net**

North America: **www.arin.net**

Note: At the time of writing, the European RIR www.ripe.net does not reply to ICMP echo requests.

The website you selected will be used in Part 2 for use with the **tracert** command.

Part 2: Trace a Route to a Remote Server Using Tracert

After you determine if your chosen websites are reachable by using **ping**, you will use **tracert** to determine the path to reach the remote server. It is helpful to look more closely at each network segment that is crossed.

Each hop in the **tracert** results displays the routes that the packets take when traveling to the final destination. The PC sends three ICMP echo request packets to the remote host. Each router in the path decrements the time to live (TTL) value by 1 before passing it onto the next system. When the decremented TTL value reaches 0, the router sends an ICMP Time Exceeded message back to the source with its IP address and the current time. When the final destination is reached, an ICMP echo reply is sent to the source host.

For example, the source host sends three ICMP echo request packets to the first hop (192.168.1.1) with the TTL value of 1. When the router 192.168.1.1 receives the echo request packets, it decrements the TTL value to 0. The router sends an ICMP Time Exceeded message back to the source. This process continues until the source hosts sends the last three ICMP echo request packets with TTL values of 8 (hop number 8 in the output below), which is the final destination. After the ICMP echo request packets arrive at the final destination, the router responds to the source with ICMP echo replies.

For hops 2 and 3, these IP addresses are private addresses. These routers are the typical setup for point-of-presence (POP) of ISP. The POP devices connect users to an ISP network.

A web-based whois tool is found at http://whois.domaintools.com/. It is used to determine the domains traveled from the source to destination.

a. At the command-line prompt, trace the route to www.cisco.com. Save the **tracert** output in a text file. Alternatively, you can redirect the output to a text file by using **>** or **>>**.

```
C:\Users\User1> tracert www.cisco.com
```

or

```
C:\Users\User1> tracert www.cisco.com > tracert-cisco.txt


Tracing route to e144.dscb.akamaiedge.net [23.67.208.170]

over a maximum of 30 hops:


    1      1 ms     <1 ms     <1 ms   192.168.1.1

    2     14 ms      7 ms      7 ms   10.39.0.1

    3     10 ms      8 ms      7 ms   172.21.0.118

    4     11 ms     11 ms     11 ms   70.169.73.196

    5     10 ms      9 ms     11 ms   70.169.75.157

    6     60 ms     49 ms        *    68.1.2.109

    7     43 ms     39 ms     38 ms   Equinix-DFW2.netarch.akamai.com [206.223.118.102]

    8     33 ms     35 ms     33 ms   a23-67-208-170.deploy.akamaitechnologies.com
[23.67.208.170]


Trace complete.
```

b. The web-based tool at http://whois.domaintools.com/ can be used to determine the owners of both the resulting IP address and domain names shown in the tracert tools output. Now perform a **tracert** to one of RIR web sites from Part 1 and save the results.

Africa: **www.afrinic.net**

Australia: **www.apnic.net**

Europe: **www.ripe.net**

South America: **www.lacnic.net**

North America: **www.arin.net**

List the domains below from your tracert results using the web-based whois tool.

_____

_____

c. Compare the lists of domains crossed to reach the final destinations.

Reflection

What can affect **tracert** results?

_____

_____

4.1.4.6 Lab – Configuring Basic Router Settings with IOS CLI

Topology

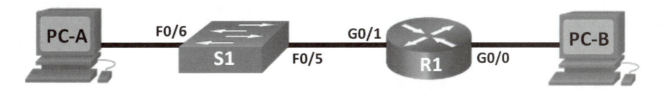

Addressing Table

Device	Interface	IP Address	Subnet Mask	Default Gateway
R1	G0/0	192.168.0.1	255.255.255.0	N/A
	G0/1	192.168.1.1	255.255.255.0	N/A
PC-A	NIC	192.168.1.3	255.255.255.0	192.168.1.1
PC-B	NIC	192.168.0.3	255.255.255.0	192.168.0.1

Objectives

Part 1: Set Up the Topology and Initialize Devices

- Cable equipment to match the network topology.

- Initialize and restart the router and switch.

Part 2: Configure Devices and Verify Connectivity

- Assign static IPv4 information to the PC interfaces.

- Configure basic router settings.

- Verify network connectivity.

- Configure the router for SSH.

Part 3: Display Router Information

- Retrieve hardware and software information from the router.

- Interpret the output from the startup configuration.

- Interpret the output from the routing table.

- Verify the status of the interfaces.

Part 4: Configure IPv6 and Verify Connectivity

Background / Scenario

This is a comprehensive lab to review previously covered IOS router commands. In Parts 1 and 2, you will cable the equipment and complete basic configurations and IPv4 interface settings on the router.

In Part 3, you will use SSH to connect to the router remotely and utilize IOS commands to retrieve information from the device to answer questions about the router. In Part 4, you will configure IPv6 on the router so that PC-B can acquire an IP address and then verify connectivity.

For review purposes, this lab provides the commands necessary for specific router configurations.

Note: The routers used with CCNA hands-on labs are Cisco 1941 Integrated Services Routers (ISRs) with Cisco IOS Release 15.2(4)M3 (universalk9 image). The switches used are Cisco Catalyst 2960 with Cisco IOS Release 15.0(2) (lanbasek9 image). Other routers, switches, and Cisco IOS versions can be used. Depending on the model and Cisco IOS version, the commands available and output produced might vary from what is shown in the labs. Refer to the Router Interface Summary Table at the end of this lab for the correct interface identifiers.

Note: Make sure that the router and switch have been erased and have no startup configurations. Refer to Appendix A for the procedures to initialize and reload devices.

Required Resources

- 1 Router (Cisco 1941 with Cisco IOS Release 15.2(4)M3 universal image or comparable)
- 1 Switch (Cisco 2960 with Cisco IOS Release 15.0(2) lanbasek9 image or comparable)
- 2 PCs (Windows 7, Vista, or XP with terminal emulation program, such as Tera Term)
- Console cables to configure the Cisco IOS devices via the console ports
- Ethernet cables as shown in the topology

Note: The Gigabit Ethernet interfaces on Cisco 1941 ISRs are autosensing and an Ethernet straight-through cable can be used between the router and PC-B. If using another model Cisco router, it may be necessary to use an Ethernet crossover cable.

Part 1: Set Up the Topology and Initialize Devices

Step 1: Cable the network as shown in the topology.

a. Attach the devices as shown in the topology diagram, and cable as necessary.

b. Power on all the devices in the topology.

Step 2: Initialize and reload the router and switch.

Note: Appendix A details the steps to initialize and reload the devices.

Part 2: Configure Devices and Verify Connectivity

Step 1: Configure the PC interfaces.

a. Configure the IP address, subnet mask, and default gateway settings on PC-A.

b. Configure the IP address, subnet mask, and default gateway settings on PC-B.

Step 2: Configure the router.

a. Console into the router and enable privileged EXEC mode.

```
Router> enable
Router#
```

b. Enter into global configuration mode.

```
Router# config terminal
Router(config)#
```

c. Assign a device name to the router.

```
Router(config)# hostname R1
```

d. Disable DNS lookup to prevent the router from attempting to translate incorrectly entered commands as though they were hostnames.

```
R1(config)# no ip domain-lookup
```

e. Require that a minimum of 10 characters be used for all passwords.

```
R1(config)# security passwords min-length 10
```

Besides setting a minimum length, list other ways to strengthen passwords.

_____

f. Assign **cisco12345** as the privileged EXEC encrypted password.

```
R1(config)# enable secret cisco12345
```

g. Assign **ciscoconpass** as the console password, establish a timeout, enable login, and add the **logging synchronous** command. The **logging synchronous** command synchronizes debug and Cisco IOS software output and prevents these messages from interrupting your keyboard input.

```
R1(config)# line con 0
R1(config-line)# password ciscoconpass
R1(config-line)# exec-timeout 5 0
R1(config-line)# login
R1(config-line)# logging synchronous
R1(config-line)# exit
R1(config)#
```

For the **exec-timeout** command, what do the **5** and **0** represent?

_____

h. Assign **ciscovtypass** as the vty password, establish a timeout, enable login, and add the **logging synchronous** command.

```
R1(config)# line vty 0 4
R1(config-line)# password ciscovtypass
R1(config-line)# exec-timeout 5 0
```

```
R1(config-line)# login
R1(config-line)# logging synchronous
R1(config-line)# exit
R1(config)#
```

i. Encrypt the clear text passwords.

```
R1(config)# service password-encryption
```

j. Create a banner that warns anyone accessing the device that unauthorized access is prohibited.

```
R1(config)# banner motd #Unauthorized access prohibited!#
```

k. Configure an IP address and interface description. Activate both interfaces on the router.

```
R1(config)# int g0/0
R1(config-if)# description Connection to PC-B
R1(config-if)# ip address 192.168.0.1 255.255.255.0
R1(config-if)# no shutdown
R1(config-if)# int g0/1
R1(config-if)# description Connection to S1
R1(config-if)# ip address 192.168.1.1 255.255.255.0
R1(config-if)# no shutdown
R1(config-if)# exit
R1(config)# exit
R1#
```

l. Set the clock on the router; for example:

```
R1# clock set 17:00:00 18 Feb 2013
```

m. Save the running configuration to the startup configuration file.

```
R1# copy running-config startup-config
Destination filename [startup-config]?
Building configuration...
[OK]
R1#
```

What would be the result of reloading the router prior to completing the **copy running-config startup-config** command?

_____

Step 3: **Verify network connectivity.**

a. Ping PC-B from a command prompt on PC-A.

Note: It may be necessary to disable the PCs firewall.

Were the pings successful? _____

After completing this series of commands, what type of remote access could be used to access R1?

_____

b. Remotely access R1 from PC-A using the Tera Term Telnet client.

Open Tera Term and enter the G0/1 interface IP address of R1 in the Host: field of the Tera Term: New Connection window. Ensure that the **Telnet** radio button is selected and then click **OK** to connect to the router.

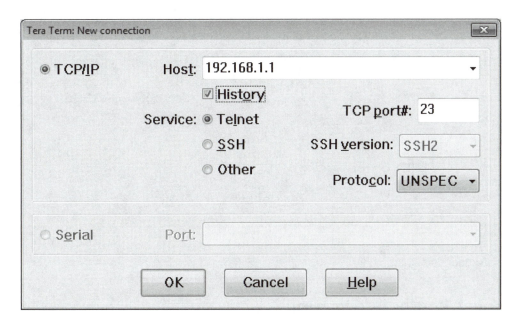

Was remote access successful? _____

Why is the Telnet protocol considered to be a security risk?

_____

Step 4: Configure the router for SSH access.

a. Enable SSH connections and create a user in the local database of the router.

```
R1# configure terminal
R1(config)# ip domain-name CCNA-lab.com
R1(config)# username admin privilege 15 secret adminpass1
R1(config)# line vty 0 4
R1(config-line)# transport input ssh
R1(config-line)# login local
R1(config-line)# exit
R1(config)# crypto key generate rsa modulus 1024
```

```
R1(config)# exit
```

b.   Remotely access R1 from PC-A using the Tera Term SSH client.

Open Tera Term and enter the G0/1 interface IP address of R1 in the Host: field of the Tera Term: New Connection window. Ensure that the **SSH** radio button is selected and then click **OK** to connect to the router.

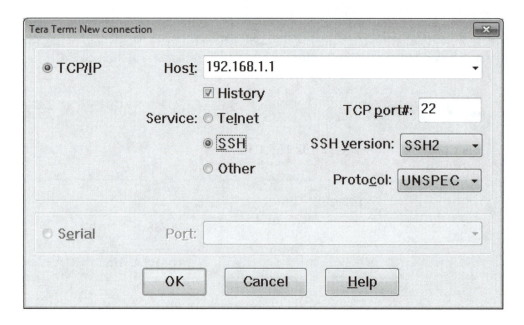

Was remote access successful? _____

Part 3:   Display Router Information

In Part 3, you will use **show** commands from an SSH session to retrieve information from the router.

Step 1:   **Establish an SSH session to R1.**

Using Tera Term on PC-B, open an SSH session to R1 at IP address 192.168.0.1 and log in as **admin** with the password **adminpass1**.

Step 2:   **Retrieve important hardware and software information.**

a.   Use the **show version** command to answer questions about the router.

What is the name of the IOS image that the router is running?

_____

How much non-volatile random-access memory (NVRAM) does the router have?

_____

How much Flash memory does the router have?

_____

_____

b. The **show** commands often provide multiple screens of outputs. Filtering the output allows a user to display certain sections of the output. To enable the filtering command, enter a pipe (|) character after a **show** command, followed by a filtering parameter and a filtering expression. You can match the output to the filtering statement by using the **include** keyword to display all lines from the output that contain the filtering expression. Filter the **show version** command, using **show version | include register** to answer the following question.

What is the boot process for the router on the next reload?

_____

Step 3:   Display the startup configuration.

Use the **show startup-config** command on the router to answer the following questions.

How are passwords presented in the output?

_____

_____

_____

Use the **show startup-config | begin vty** command.

What is the result of using this command?

_____

_____

Step 4:   **Display the routing table on the router.**

Use the **show ip route** command on the router to answer the following questions.

What code is used in the routing table to indicate a directly connected network?

_____

_____

How many route entries are coded with a C code in the routing table? _____

Step 5: Display a summary list of the interfaces on the router.

Use the **show ip interface brief** command on the router to answer the following question.

What command changed the status of the Gigabit Ethernet ports from administratively down to up?

_____

Part 4: Configure IPv6 and Verify Connectivity

Step 1: Assign IPv6 addresses to R1 G0/0 and enable IPv6 routing.

Note: Assigning an IPv6 address in addition to an IPv4 address on an interface is known as dual stacking, because both the IPv4 and IPv6 protocol stacks are active. By enabling IPv6 unicast routing on R1, PC-B receives the R1 G0/0 IPv6 network prefix and can autoconfigure its IPv6 address and its default gateway.

a. Assign an IPv6 global unicast address to interface G0/0, assign the link-local address in addition to the unicast address on the interface, and enable IPv6 routing.

```
R1# configure terminal
R1(config)# interface g0/0
R1(config-if)# ipv6 address 2001:db8:acad:a::1/64
R1(config-if)# ipv6 address fe80::1 link-local
R1(config-if)# no shutdown
R1(config-if)# exit
R1(config)# ipv6 unicast-routing
R1(config)# exit
```

b. Use the **show ipv6 int brief** command to verify IPv6 settings on R1.

If no IPv6 address is assigned to G0/1, why is it listed as [up/up]?

_____

_____

c. Issue the **ipconfig** command on PC-B to examine the IPv6 configuration.

What is the IPv6 address assigned to PC-B?

_____

What is the default gateway assigned to PC-B? _____

Issue a ping from PC-B to the R1 default gateway link local address. Was it successful? _____

Issue a ping from PC-B to the R1 IPv6 unicast address 2001:db8:acad:a::1. Was it successful? _____

Reflection

1. In researching a network connectivity issue, a technician suspects that an interface was not enabled. What **show** command could the technician use to troubleshoot this issue?

_____

2. In researching a network connectivity issue, a technician suspects that an interface was assigned an incorrect subnet mask. What **show** command could the technician use to troubleshoot this issue?

_____

3. After configuring IPv6 on the R1 G0/0 PC-B LAN, if you were to ping from PC-A to the PC-B IPv6 address, would the ping succeed? Why or why not?

_____

Router Interface Summary Table

Router Interface Summary				
Router Model	**Ethernet Interface #1**	**Ethernet Interface #2**	**Serial Interface #1**	**Serial Interface #2**
1800	Fast Ethernet 0/0 (F0/0)	Fast Ethernet 0/1 (F0/1)	Serial 0/0/0 (S0/0/0)	Serial 0/0/1 (S0/0/1)
1900	Gigabit Ethernet 0/0 (G0/0)	Gigabit Ethernet 0/1 (G0/1)	Serial 0/0/0 (S0/0/0)	Serial 0/0/1 (S0/0/1)
2801	Fast Ethernet 0/0 (F0/0)	Fast Ethernet 0/1 (F0/1)	Serial 0/1/0 (S0/1/0)	Serial 0/1/1 (S0/1/1)
2811	Fast Ethernet 0/0 (F0/0)	Fast Ethernet 0/1 (F0/1)	Serial 0/0/0 (S0/0/0)	Serial 0/0/1 (S0/0/1)
2900	Gigabit Ethernet 0/0 (G0/0)	Gigabit Ethernet 0/1 (G0/1)	Serial 0/0/0 (S0/0/0)	Serial 0/0/1 (S0/0/1)

Note: To find out how the router is configured, look at the interfaces to identify the type of router and how many interfaces the router has. There is no way to effectively list all the combinations of configurations for each router class. This table includes identifiers for the possible combinations of Ethernet and Serial interfaces in the device. The table does not include any other type of interface, even though a specific router may contain one. An example of this might be an ISDN BRI interface. The string in parenthesis is the legal abbreviation that can be used in Cisco IOS commands to represent the interface.

Appendix A: Initializing and Reloading a Router and Switch

Step 1:   Initialize and reload the router.

a. Console into the router and enable privileged EXEC mode.

```
Router> enable
Router#
```

b. Type the **erase startup-config** command to remove the startup configuration from NVRAM.

```
Router# erase startup-config
Erasing the nvram filesystem will remove all configuration files! Continue? [confirm]
[OK]
Erase of nvram: complete
Router#
```

c. Issue the **reload** command to remove an old configuration from memory. When prompted to **Proceed with reload**, press Enter to confirm the reload. (Pressing any other key aborts the reload.)

```
Router# reload

Proceed with reload? [confirm]

*Nov 29 18:28:09.923: %SYS-5-RELOAD: Reload requested by console. Reload Reason: Re-
load Command.
```

Note: You may be prompted to save the running configuration prior to reloading the router. Type **no** and press Enter.

```
System configuration has been modified. Save? [yes/no]: no
```

d. After the router reloads, you are prompted to enter the initial configuration dialog. Enter **no** and press Enter.

```
Would you like to enter the initial configuration dialog? [yes/no]: no
```

e. You are prompted to terminate autoinstall. Type **yes** and then press Enter.

```
Would you like to terminate autoinstall? [yes]: yes
```

Step 2:   Initialize and reload the switch.

a. Console into the switch and enter privileged EXEC mode.

```
Switch> enable
Switch#
```

b. Use the **show flash** command to determine if any VLANs have been created on the switch.

```
Switch# show flash

Directory of flash:/

    2  -rwx        1919   Mar 1 1993 00:06:33 +00:00  private-config.text
    3  -rwx        1632   Mar 1 1993 00:06:33 +00:00  config.text
    4  -rwx       13336   Mar 1 1993 00:06:33 +00:00  multiple-fs
    5  -rwx    11607161   Mar 1 1993 02:37:06 +00:00  c2960-lanbasek9-mz.150-2.SE.bin
    6  -rwx         616   Mar 1 1993 00:07:13 +00:00  vlan.dat

32514048 bytes total (20886528 bytes free)
Switch#
```

c. If the **vlan.dat** file was found in flash, then delete this file.

```
Switch# delete vlan.dat

Delete filename [vlan.dat]?
```

d. You are prompted to verify the filename. At this point, you can change the filename or just press Enter if you have entered the name correctly.

e. You are prompted to confirm deleting this file. Press Enter to confirm deletion. (Pressing any other key aborts the deletion.)

```
Delete flash:/vlan.dat? [confirm]
Switch#
```

f.   Use the **erase startup-config** command to erase the startup configuration file from NVRAM. You are prompted to confirm removing the configuration file. Press Enter to confirm to erase this file. (Pressing any other key aborts the operation.)

```
Switch# erase startup-config
Erasing the nvram filesystem will remove all configuration files! Continue? [confirm]
[OK]
Erase of nvram: complete
Switch#
```

g.   Reload the switch to remove any old configuration information from memory. You are prompted to confirm reloading the switch. Press Enter to proceed with the reload. (Pressing any other key aborts the reload.)

```
Switch# reload
Proceed with reload? [confirm]
```

Note: You may be prompted to save the running configuration prior to reloading the switch. Type **no** and press Enter.

```
System configuration has been modified. Save? [yes/no]: no
```

h.   After the switch reloads, you should be prompted to enter the initial configuration dialog. Type **no** and press Enter.

```
Would you like to enter the initial configuration dialog? [yes/no]: no
Switch>
```

4.1.4.7 Lab – Configuring Basic Router Settings with CCP

Topology

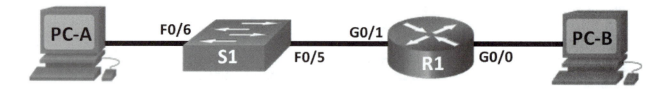

Addressing Table

Device	Interface	IP Address	Subnet Mask	Default Gateway
R1	G0/0	192.168.0.1	255.255.255.0	N/A
	G0/1	192.168.1.1	255.255.255.0	N/A
S1	VLAN 1	N/A	N/A	N/A
PC-A	NIC	192.168.1.3	255.255.255.0	192.168.1.1
PC-B	NIC	192.168.0.3	255.255.255.0	192.168.0.1

Objectives

Part 1: Set Up the Topology and Initialize Devices

Part 2: Configure Devices and Verify Connectivity

Part 3: Configure Router to Allow CCP Access

Part 4: (Optional) Install and Set Up CCP on PC-A

Part 5: Configure R1 Settings Using CCP

Part 6: Use CCP Utilities

Background / Scenario

Cisco Configuration Professional (CCP) is a PC-based application that provides GUI-based device management for Integrated Services Routers (ISRs). It simplifies the configuration of routing, firewall, VPN, WAN, LAN, and other settings through menus and easy-to-use wizards.

In this lab, you will configure the router settings using the configuration from the previous lab in this chapter. Layer 3 connectivity must be established between the PC running CCP (PC-A) and R1 before CCP can establish a connection. In addition, HTTP access and authentication must be configured on R1.

You will download and install CCP on the PC and then use it to monitor R1's interface status, configure an interface, set the date and time, add a user to the local database, and edit vty settings. You will also use some of the utilities included in CCP.

Note: Router configurations performed using CCP generate IOS CLI commands. CCP can be very useful for configuring more complex router features because it does not require specific knowledge of the Cisco IOS command syntax.

Note: The routers used with CCNA hands-on labs are Cisco 1941 Integrated Services Routers (ISRs) with Cisco IOS Release 15.2(4)M3 (universalk9 image). The switches used are Cisco Catalyst 2960s with Cisco IOS Release 15.0(2) (lanbasek9 image). Other routers, switches, and Cisco IOS versions can be used. Depending on the model and Cisco IOS version, the commands available and output produced might vary from what is shown in the labs. Refer to the Router Interface Summary Table at the end of this lab for the correct interface identifiers.

Note: Make sure that the router and switch have been erased and have no startup configurations. If you are unsure, contact your instructor.

Required Resources

- 1 Router (Cisco 1941 with Cisco IOS Release 15.2(4)M3 universal image or comparable)
- 1 Switch (Cisco 2960 with Cisco IOS Release 15.0(2) lanbasek9 image or comparable)
- 2 PCs (Windows 7, Vista, or XP with terminal emulation program, such as Tera Term)
- Console cables to configure the Cisco IOS devices via the console ports
- Ethernet cables as shown in the topology

Note: PC system requirements for CCP version 2.6 are:

- 2 GHz processor or faster
- 1 GB DRAM minimum; 2 GB recommended
- 400 MB of available hard disk space
- Internet Explorer 6.0 or above
- Screen resolution of 1024x768 or higher
- Java Runtime Environment (JRE) version 1.6.0_11 or later.
- Adobe Flash Player version 10.0 or later, with Debug set to No

Note: The Gigabit Ethernet interfaces on Cisco 1941 ISRs are autosensing and an Ethernet straight-through cable may be used between the router and PC-B. If using another model Cisco router, it may be necessary to use an Ethernet crossover cable.

Part 1:   Set Up the Topology and Initialize Devices

Step 1:   Cable the network as shown in the topology.

a.   Attach the devices shown in the topology diagram, and cable as necessary.

b.   Power on all the devices in the topology.

Step 2:   Initialize and reload the router and switch.

Part 2:   Configure Devices and Verify Connectivity

In Part 2, you will configure basic settings, such as the interface IP addresses (G0/1 only), secure device access, and passwords. Refer to the Topology and Addressing Table for device names and address information.

Step 1:   **Configure the PC interfaces.**

a.   Configure the IP address, subnet mask, and default gateway settings on PC-A.

b.   Configure the IP address, subnet mask, and default gateway settings on PC-B.

Step 2:   **Configure the router.**

Note: Do NOT configure interface G0/0 at this time. You will configure this interface using CCP later in the lab.

a.   Console into the router and enable privileged EXEC mode.

b.   Enter into global configuration mode.

c.   Disable DNS lookup.

d.   Assign a device name to the router.

e.   Require that a minimum of 10 characters be used for all passwords.

f.   Assign **cisco12345** as the privileged EXEC encrypted password.

g.   Assign **ciscoconpass** as the console password and enable login.

h.   Assign **ciscovtypass** as the vty password and enable login,

i.   Configure **logging synchronous** on the console and vty lines.

j.   Encrypt the clear text passwords.

k.   Create a banner that warns anyone accessing the device that unauthorized access is prohibited.

l.   Configure the IP addresses, an interface description, and activate G0/1 interface on the router.

m. Save the running configuration to the startup configuration file.

Step 3:   **Verify network connectivity.**

Verify that you can ping R1 G0/1 from PC-A.

Part 3:   Configure the Router to Allow CCP Access

In Part 3, you will set up the router to allow CCP access by enabling HTTP and HTTPS server services. You will also enable HTTP authentication to use the local database.

Step 1:   **Enable HTTP and HTTPS server services on the router.**

```
R1(config)# ip http server
R1(config)# ip http secure-server
```

Step 2:   **Enable HTTP authentication to use the local database on the router.**

```
R1(config)# ip http authentication local
```

Step 3:   **Configure the router for CCP access.**

Assign a user in the router local database for accessing CCP using username **admin** and password **admin-pass1**.

```
R1(config)# username admin privilege 15 secret adminpass1
```

Part 4:   (Optional) Install and Set Up CCP on PC-A

Step 1:   **Install CCP.**

Note: This step can be skipped if CCP is already installed on PC-A.

a.   Download CCP 2.6 from Cisco's website:

http://software.cisco.com/download/release.html?mdfid=281795035&softwareid=282159854&release=2.6 &rellifecycle=&relind=AVAILABLE&reltype=all

b.   Choose the **cisco-config-pro-k9-pkg-2_6-en.zip** file.

Note: Verify that you select the correct CCP file and not CCP Express. If there is a more current release of CCP, you may choose to download it; however, this lab is based on CCP 2.6.

c.   Agree to the terms and conditions, and download and save the file to the desired location.

d.   Open the zip file and run the CCP executable.

e.   Follow the on-screen instructions to install CCP 2.6 on your PC.

Step 2:   **Change settings to run as the administrator.**

CCP may fail to launch correctly if it is not run as an administrator. You can change the launch settings so that it automatically runs in administrator mode.

a.   Right-click the **CCP** desktop icon (or click the **Start** button) and then right-click **Cisco Configuration Professional**. In the drop-down list, select **Properties**.

b.   In the Properties dialog box, select the **Compatibility** tab. In the Privilege Level section, click the **Run this program as an administrator** checkbox, and then click **OK**.

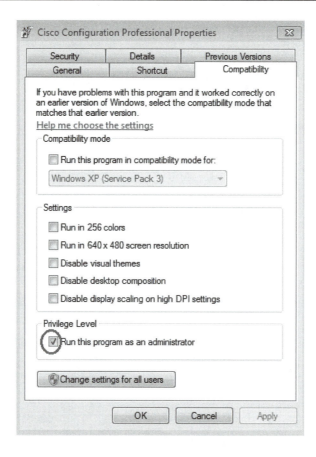

Step 3: **Create or manage communities.**

a. On PC-A, start CCP. (Double-click the CCP desktop icon or click **Start** > **Cisco Configuration Professional**.)

b. If you receive a security warning message prompting to allow the CiscoCP.exe program to make changes to the computer, click **Yes**.

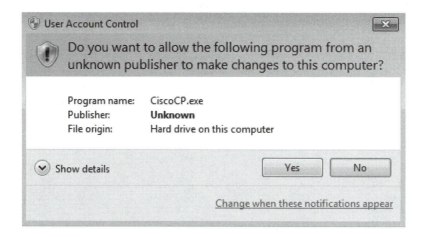

c. When CCP starts, the **Select / Mange Community** dialog box displays. Enter the IP address for R1 G0/1, and the username **admin** and password **adminpass1** that you added to the local database during the router configuration in Part 2. Click **OK**.

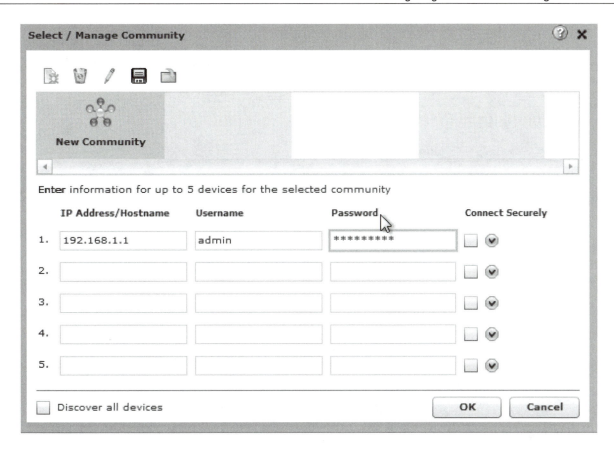

d. In the Community Information window, click **Discover**.

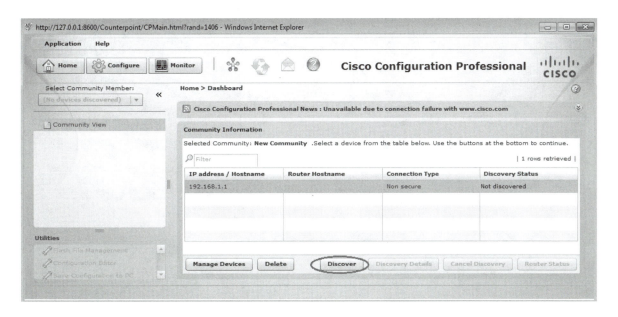

If you have configured the router correctly, the Discovery Status changes from **Not discovered** to **Discovered**, and R1 appears in the Router Hostname column.

Note: If there is a problem with your configuration, you will see a "Discovery failed" status. Click **Discovery Details** to determine why the discovery process failed and then troubleshoot the problem.

Part 5:   Configure R1 Settings Using CCP

In Part 5, you will use CCP to display information about R1, configure interface G0/0, set the date and time, add a user to the local database, and change your vty settings.

Step 1:   View the status of the interfaces on R1.

a.   On the CCP toolbar, click **Monitor**.

b.   In the left navigation pane, click **Router** > **Overview** to display the Monitor Overview screen in the right content pane.

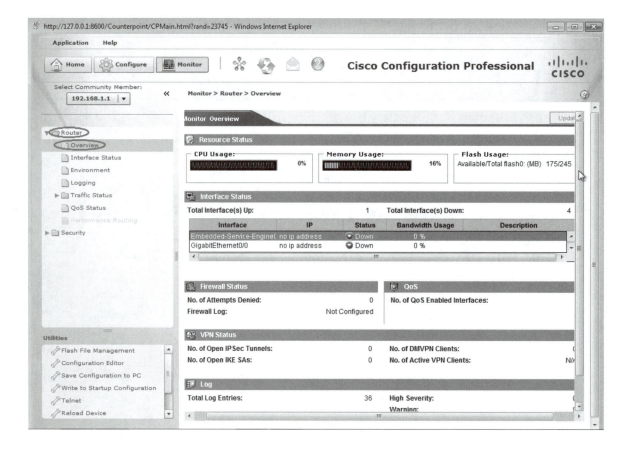

c.   Use the up and down arrows to the right of the interface list to scroll through the list of interfaces for the router.

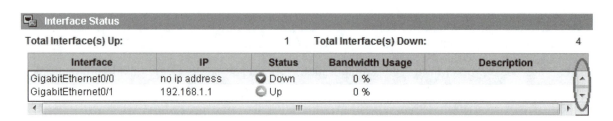

Step 2:   Use the Ethernet LAN wizard to configure interface G0/0.

a.   On the CCP toolbar, click **Configure**.

b.   In the left navigation pane, click **Interface Management** > **Interface and Connections** to display the Interfaces and Connections screen in the right content pane.

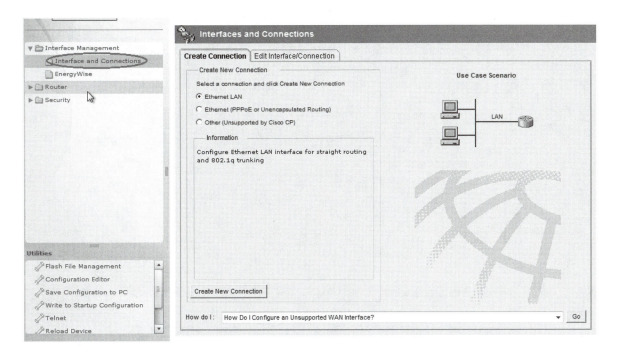

c.   Click **Create New Connection** to start the Ethernet LAN wizard.

d.   When you are prompted to enable AAA on the router, click **No**.

e.   Click **Next** to be guided through the Layer 3 Ethernet interface creation process.

f.   Keep the **Configure this interface for straight routing** radio button selected and click **Next**.

g.   Enter **192.168.0.1** in the IP address field and **255.255.255.0** in the Subnet mask field and click **Next**.

h.   Keep the **No** radio button selected on the DHCP server screen and click **Next**.

i.   Review the summary screen and click **Finish**.

j.   Click the **Save running config to device's startup config** check box, and then click **Deliver**. This adds the commands shown in the preview window to the running configuration, and then saves the running configuration to the startup configuration on the router.

k.   The Commands Delivery Status window displays. Click **OK** to close this window. You will be routed back to the Interfaces and Connections screen; G0/0 should have turned green and displayed as Up in the Status column.

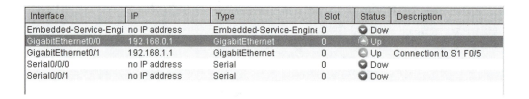

Step 3:   Set the date and time on the router.

a. In the left navigation pane, select **Router** > **Time** > **Date and Time** to display the Additional Tasks > Date/Time screen in the right content pane. Click **Change Settings…**.

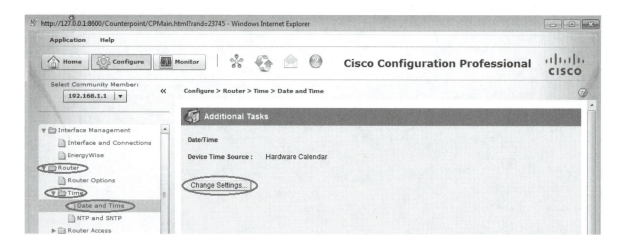

b. In the Date and Time Properties window, edit the Date, Time, and Time Zone. Click **Apply.**

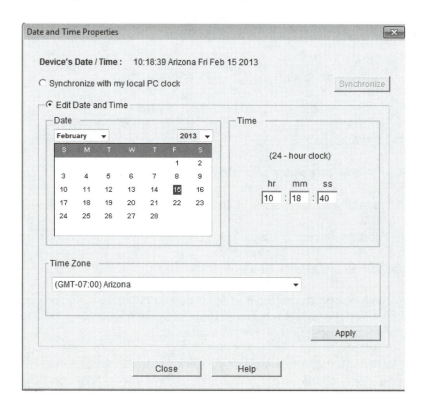

c. In the Router's clock configured window, click **OK**. In the Date and Time Properties window, click **Close**.

Step 4:   Add a new user account to the local database.

a. In the left navigation pane, select **Router** > **Router Access** > **User Accounts/View** to display the Additional Tasks > User Accounts/View screen in the content pane on the right. Click the **Add...** button.

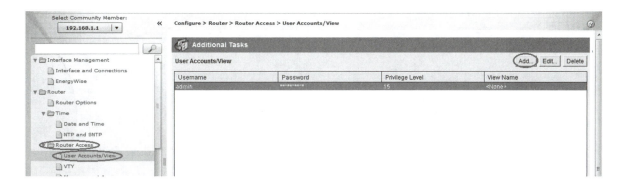

b. Enter **ccpadmin** in the Username: field. Enter **ciscoccppass** in the New Password: and Confirm New Password: fields. Select **15** in the Privilege Level: drop-down list. Click **OK** to add this user to the local database.

c. In the Deliver Configuration to Device window, click the **Save running config to device's startup config** check box, and then click **Deliver**.

d. Review the information in the Commands Delivery Status window, and click **OK**. The new user account should now appear in the content pane on the right.

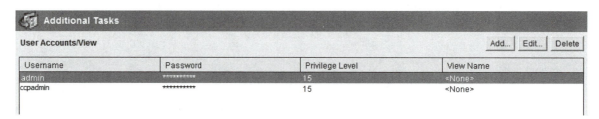

Step 5:   **Edit vty line settings.**

a. In the left navigation pane, select **Router Access** > **VTY** to display the Additional Tasks > VTYs screen in the content pane on the right. Click **Edit....**

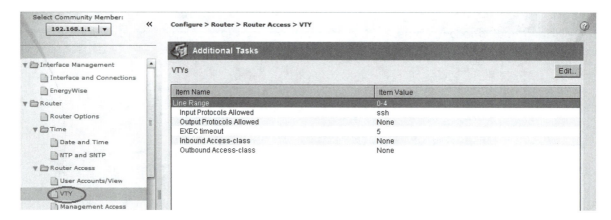

b. In the Edit VTY Lines window, change the Time out: field to **15** minutes. Click the **Input Protocol** > **Telne**t check box. Review the other options available. Also select the **SSH** checkbox. Then click **OK**.

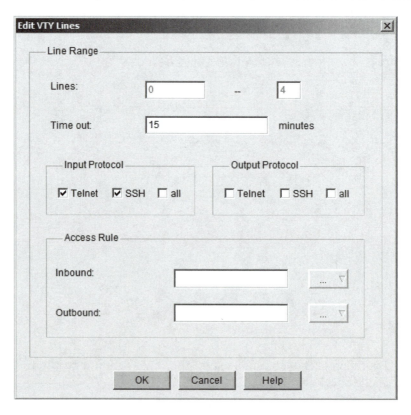

c. Review the commands that will be delivered to the running configuration on the Deliver Configuration to Device screen and click **Deliver**. In the Commands Delivery Status window, click **OK**. The content pane on the right should reflect the changes to the EXEC timeout value.

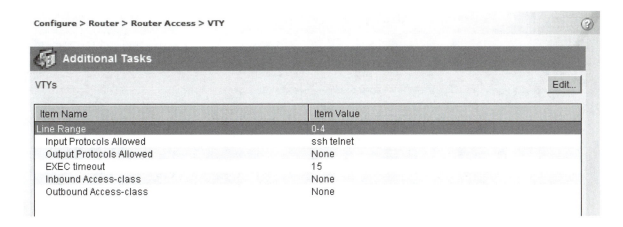

Part 6: Use CCP Utilities

In Part 6, you will use the Utilities pane to save the router's running configuration to the startup configuration. The Ping utility will be used to test network connectivity, and the View utility will be used to show the router's running configuration. Finally, you will close CCP.

Step 1:   Save the router's running configuration to the startup configuration.

a. At the bottom of the left navigation pane, locate the Utilities pane. Click **Write to Startup Configuration**.

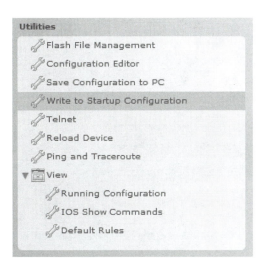

b. The content pane displays a confirmation screen. Click **Confirm**. An Information window displays, letting you know that the configuration was saved successfully. Click **OK**.

Step 2:   Use the Ping utility to test connectivity to PC-B.

a. In the Utilities pane, click **Ping and Traceroute** to display the Ping and Traceroute screen in the content pane. Enter **192.168.0.3** in the Destination*: field and then click **Ping**. Use the scrollbar to the right of the results box to view the results of your ping.

Step 3: **Use the View utility to show the running configuration for the router.**

a. In the Utilities pane, click **View** > **IOS Show Commands** to display the IOS Show Commands screen in the content pane.

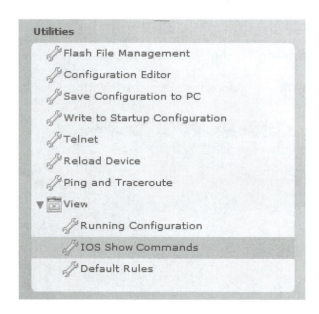

b. Select **show run** from the drop-down list and click **Show**. The router's running configuration is displayed in the content pane.

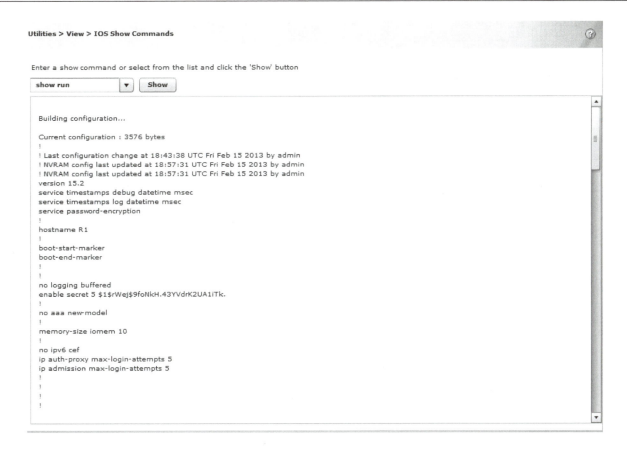

Step 4:   Close CCP.

Close the CCP window. When a Windows Internet Explorer confirmation window displays, click **Leave this page**.

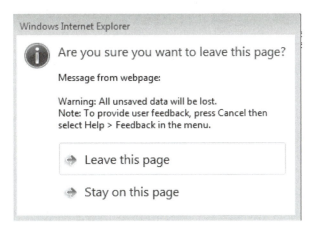

Reflection

1. What transport protocol does CCP use to access the router and what commands are used to allow access?

_____

_____

2. What router command tells CCP to use the local database to authenticate?

_____

3. What other **show** commands are available in the Utilities pane of CCP?

_____

_____

4. Why would you want to use CCP instead of the IOS CLI?

_____

_____

_____

Router Interface Summary Table

Router Interface Summary				
Router Model	**Ethernet Interface #1**	**Ethernet Interface #2**	**Serial Interface #1**	**Serial Interface #2**
1800	Fast Ethernet 0/0 (F0/0)	Fast Ethernet 0/1 (F0/1)	Serial 0/0/0 (S0/0/0)	Serial 0/0/1 (S0/0/1)
1900	Gigabit Ethernet 0/0 (G0/0)	Gigabit Ethernet 0/1 (G0/1)	Serial 0/0/0 (S0/0/0)	Serial 0/0/1 (S0/0/1)
2801	Fast Ethernet 0/0 (F0/0)	Fast Ethernet 0/1 (F0/1)	Serial 0/1/0 (S0/1/0)	Serial 0/1/1 (S0/1/1)
2811	Fast Ethernet 0/0 (F0/0)	Fast Ethernet 0/1 (F0/1)	Serial 0/0/0 (S0/0/0)	Serial 0/0/1 (S0/0/1)
2900	Gigabit Ethernet 0/0 (G0/0)	Gigabit Ethernet 0/1 (G0/1)	Serial 0/0/0 (S0/0/0)	Serial 0/0/1 (S0/0/1)
Note: To find out how the router is configured, look at the interfaces to identify the type of router and how many interfaces the router has. There is no way to effectively list all the combinations of configurations for each router class. This table includes identifiers for the possible combinations of Ethernet and Serial interfaces in the device. The table does not include any other type of interface, even though a specific router may contain one. An example of this might be an ISDN BRI interface. The string in parenthesis is the legal abbreviation that can be used in Cisco IOS commands to represent the interface.				

4.4.1.1 Class Activity – We Really Could Use a Map!

Objectives

Describe the three types of routes that are populated in a routing table (to include: directly-connected, static, and dynamic).

Scenario

Use the Ashland and Richmond routing tables shown below. With the help of a classmate, draw a network topology using the information from the tables. To assist you with this activity, follow these guidelines:

- Start with the Ashland router - use its routing table to identify ports and IP addresses/networks.
- Add the Richmond router - use its routing table to identify ports and IP addresses/networks.
- Add any other intermediary and end devices, as specified by the tables.

In addition, record answers from your group to the reflection questions provided with this activity.

Be prepared to share your work with another group or the class.

Resources

```
Ashland> show ip route
Codes: L - local, C - connected, S - static, R - RIP, M - mobile, B - BGP
D - EIGRP, EX - EIGRP external, O - OSPF, IA - OSPF inter area
N1 - OSPF NSSA external type 1, N2 - OSPF NSSA external type 2
E1 - OSPF external type 1, E2 - OSPF external type 2, E - EGP
i - IS-IS, L1 - IS-IS level-1, L2 - IS-IS level-2, ia - IS-IS inter area
* - candidate default, U - per-user static route, o - ODR
P - periodic downloaded static route

Gateway of last resort is not set

     192.168.1.0/24 is variably subnetted, 2 subnets, 2 masks
C    192.168.1.0/24 is directly connected, GigabitEthernet0/1
L    192.168.1.1/32 is directly connected, GigabitEthernet0/1
     192.168.2.0/24 is variably subnetted, 2 subnets, 2 masks
C    192.168.2.0/24 is directly connected, Serial0/0/0
L    192.168.2.1/32 is directly connected, Serial0/0/0
D    192.168.3.0/24 [90/2170368] via 192.168.4.2, 01:53:50, GigabitEthernet0/0
     192.168.4.0/24 is variably subnetted, 2 subnets, 2 masks
C    192.168.4.0/24 is directly connected, GigabitEthernet0/0
L    192.168.4.1/32 is directly connected, GigabitEthernet0/0
D    192.168.5.0/24 [90/3072] via 192.168.4.2, 01:59:14, GigabitEthernet0/0
S    192.168.6.0/24 [1/0] via 192.168.2.2
Ashland>
```

```
Richmond> show ip route
Codes: L - local, C - connected, S - static, R - RIP, M - mobile, B - BGP
       D - EIGRP, EX - EIGRP external, O - OSPF, IA - OSPF inter area
       N1 - OSPF NSSA external type 1, N2 - OSPF NSSA external type 2
       E1 - OSPF external type 1, E2 - OSPF external type 2, E - EGP
       i - IS-IS, L1 - IS-IS level-1, L2 - IS-IS level-2, ia - IS-IS inter area
       * - candidate default, U - per-user static route, o - ODR
       P - periodic downloaded static route

Gateway of last resort is not set

S    192.168.1.0/24 [1/0] via 192.168.3.1
D    192.168.2.0/24 [90/2170368] via 192.168.5.2, 01:55:09, GigabitEthernet0/1
     192.168.3.0/24 is variably subnetted, 2 subnets, 2 masks
C    192.168.3.0/24 is directly connected, Serial0/0/0
L    192.168.3.2/32 is directly connected, Serial0/0/0
D    192.168.4.0/24 [90/3072] via 192.168.5.2, 01:55:09, GigabitEthernet0/1
     192.168.5.0/24 is variably subnetted, 2 subnets, 2 masks
C    192.168.5.0/24 is directly connected, GigabitEthernet0/1
L    192.168.5.1/32 is directly connected, GigabitEthernet0/1
     192.168.6.0/24 is variably subnetted, 2 subnets, 2 masks
C    192.168.6.0/24 is directly connected, GigabitEthernet0/0
L    192.168.6.1/32 is directly connected, GigabitEthernet0/0
Richmond>
```

Reflection

1. How many directly connected routes are listed on the Ashland router? What letter represents a direct connection to a network on a routing table?

2. Find the route to the 192.168.6.0/24 network. What kind of route is this? Was it dynamically discovered by the Ashland router or manually configured by a network administrator on the Ashland router?

3. If you were configuring a default (static route) to any network from the Ashland router and wanted to send all data to 192.168.2.2 (the next hop) for routing purposes, how would you write it?

4. If you were configuring a default (static route) to any network from the Ashland router and wanted to send all data through your exit interface, how would you write it?

5. When would you choose to use static routing, instead of letting dynamic routing take care of the routing paths for you?

6. What is the significance of the L on the left side of the routing table?

Chapter 5 — Inter-VLAN Routing

5.0.1.2 Class Activity – Switching to Local-Network Channels

Objective

Configure routing between VLANs in a small to medium-sized business network.

Scenario

You work for a small- to medium-sized business. As the network administrator, you are responsible for ensuring that your network operates efficiently and securely.

Several years ago, you created VLANs on your only switch for two of your departments, Accounting and Sales. As the business has grown, it has become apparent that sometimes these two departments must share company files and network resources.

You discuss this scenario with network administrators in a few branches of your company. They tell you to consider using inter-VLAN routing.

Research the concept of inter-VLAN routing. Design a simple presentation to show your manager how you would use inter-VLAN routing to allow the Accounting and Sales departments to remain separate, but share company files and network resources.

Resources

- Internet connection
- Software presentation program

Directions

Work with a partner to complete this activity.

Step 1:   **Use your Internet connection to research how inter-VLANs operate.**

    a.   Use a search engine to locate a few basic articles, or short videos, that discuss the concept of inter-VLAN routing.

    b.   Read the articles, or view the videos, and take notes about how VLANs operate.

    c.   Make sure you record where the information was found so that you can include the sources in Step 2 of this activity.

Step 2:   **Create a presentation for your manager.**

    a.   Design a small presentation for your manager listing how you would set up an inter-VLAN routing-based network for your small- to medium-sized business, using no more than five slides.

b. Include slides which focus on:

1) A synopsis of reasons you would change your current network to an inter-VLAN-switched network. Restate what you are trying to accomplish in your design proposal.

2) A basic, easily understood definition and benefits of using inter-VLAN routing.

3) A graphic depicting how you would modify your current network to use inter-VLAN routing.

   a) Your current network utilizes one Cisco 2960 switch and one Cisco 1941 series router.

   b) Funding for new equipment is not negotiable.

4) How inter-VLANs could continue to assist with network traffic yet allow departments to communicate with each other.

5) How inter-VLAN routing would scale for the future.

c. Make sure you quote the sources upon which you are basing your presentation.

Step 3: **Present your proposal to the entire class.**

5.1.2.4 Lab – Configuring Per-Interface Inter-VLAN Routing

Topology

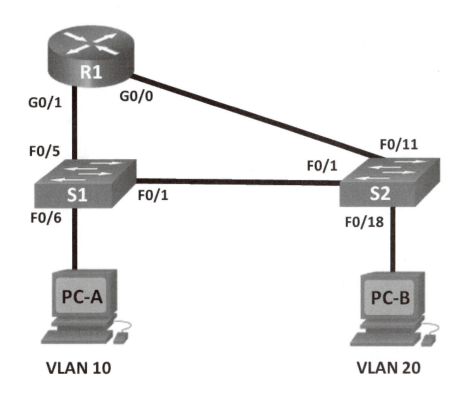

Addressing Table

Device	Interface	IP Address	Subnet Mask	Default Gateway
R1	G0/0	192.168.20.1	255.255.255.0	N/A
	G0/1	192.168.10.1	255.255.255.0	N/A
S1	VLAN 10	192.168.10.11	255.255.255.0	192.168.10.1
S2	VLAN 10	192.168.10.12	255.255.255.0	192.168.10.1
PC-A	NIC	192.168.10.3	255.255.255.0	192.168.10.1
PC-B	NIC	192.168.20.3	255.255.255.0	192.168.20.1

Objectives

Part 1: Build the Network and Configure Basic Device Settings

Part 2: Configure Switches with VLANs and Trunking

Part 3: Verify Trunking, VLANs, Routing, and Connectivity

Background / Scenario

Legacy inter-VLAN routing is seldom used in today's networks; however, it is helpful to configure and understand this type of routing before moving on to router-on-a-stick (trunk-based) inter-VLAN routing or configuring Layer-3 switching. Also, you may encounter per-interface inter-VLAN routing in organizations with very small networks. One of the benefits of legacy inter-VLAN routing is ease of configuration.

In this lab, you will set up one router with two switches attached via the router Gigabit Ethernet interfaces. Two separate VLANs will be configured on the switches, and you will set up routing between the VLANs.

Note: This lab provides minimal assistance with the actual commands necessary to configure the router and switches. The required switch VLAN configuration commands are provided in Appendix A of this lab. Test your knowledge by trying to configure the devices without referring to the appendix.

Note: The routers used with CCNA hands-on labs are Cisco 1941 Integrated Services Routers (ISRs) with Cisco IOS, Release 15.2(4)M3 (universalk9 image). The switches used are Cisco Catalyst 2960s with Cisco IOS, Release 15.0(2) (lanbasek9 image). Other routers, switches and Cisco IOS versions can be used. Depending on the model and Cisco IOS version, the commands available and output produced might vary from what is shown in the labs. Refer to the Router Interface Summary Table at the end of this lab for the correct interface identifiers.

Note: Make sure that the routers and switches have been erased and have no startup configurations. If you are unsure, contact your instructor.

Required Resources

- 1 Router (Cisco 1941 with Cisco IOS Release 15.2(4)M3 universal image or comparable)
- 2 Switches (Cisco 2960 with Cisco IOS Release 15.0(2) lanbasek9 image or comparable)
- 2 PCs (Windows 7, Vista, or XP with terminal emulation program, such as Tera Term)
- Console cables to configure the Cisco IOS devices via the console ports
- Ethernet cables as shown in the topology

Part 1: Build the Network and Configure Basic Device Settings

In Part 1, you will set up the network topology and clear any configurations, if necessary.

Step 1: Cable the network as shown in the topology.

Step 2: Initialize and reload the router and switches.

Step 3: Configure basic settings for R1.

a. Disable DNS lookup.

b. Assign the device name.

c. Assign **class** as the privileged EXEC mode encrypted password.

d. Assign **cisco** as the console and vty line password and enable login.

e. Configure addressing on G0/0 and G0/1 and enable both interfaces.

Step 4:   **Configure basic settings on S1 and S2.**

a.   Disable DNS lookup.

b.   Assign the device name.

c.   Assign **class** as the privileged EXEC mode encrypted password.

d.   Assign **cisco** as the console and vty line password and enable login.

Step 5:   **Configure basic settings on PC-A and PC-B.**

Configure PC-A and PC-B with IP addresses and a default gateway address according to the Addressing Table.

Part 2:   Configure Switches with VLANs and Trunking

In Part 2, you will configure the switches with VLANs and trunking.

Step 1:   **Configure VLANs on S1.**

a.   On S1, create VLAN 10. Assign **Student** as the VLAN name.

b.   Create VLAN 20. Assign **Faculty-Admin** as the VLAN name.

c.   Configure F0/1 as a trunk port.

d.   Assign ports F0/5 and F0/6 to VLAN 10 and configure both F0/5 and F0/6 as access ports.

e.   Assign an IP address to VLAN 10 and enable it. Refer to the Addressing Table.

f.   Configure the default gateway according to the Addressing Table.

Step 2:   **Configure VLANs on S2.**

a.   On S2, create VLAN 10. Assign **Student** as the VLAN name.

b.   Create VLAN 20. Assign **Faculty-Admin** as the VLAN name.

c.   Configure F0/1 as a trunk port.

d.   Assign ports F0/11 and F0/18 to VLAN 20 and configure both F0/11 and F0/18 as access ports.

e.   Assign an IP address to VLAN 10 and enable it. Refer to the Addressing Table.

f.   Configure the default gateway according to the Addressing Table.

Part 3:   Verify Trunking, VLANs, Routing, and Connectivity

Step 1:   **Verify the R1 routing table.**

a.   On R1, issue the **show ip route** command. What routes are listed on R1?

_____

_____

b. On both S1 and S2, issue the **show interface trunk** command. Is the F0/1 port on both switches set to trunk? _____

c. Issue a **show vlan brief** command on both S1 and S2. Verify that VLANs 10 and 20 are active and that the proper ports on the switches are in the correct VLANs. Why is F0/1 not listed in any of the active VLANs?

_____

d. Ping from PC-A in VLAN 10 to PC-B in VLAN 20. If Inter-VLAN routing is functioning correctly, the pings between the 192.168.10.0 network and the 192.168.20.0 should be successful.

   Note: It may be necessary to disable the PC firewall to ping between PCs.

e. Verify connectivity between devices. You should be able to ping between all devices. Troubleshoot if you are not successful.

Reflection

What is an advantage of using legacy inter-VLAN routing?

_____

_____

_____

Router Interface Summary Table

Router Interface Summary				
Router Model	**Ethernet Interface #1**	**Ethernet Interface #2**	**Serial Interface #1**	**Serial Interface #2**
1800	Fast Ethernet 0/0 (F0/0)	Fast Ethernet 0/1 (F0/1)	Serial 0/0/0 (S0/0/0)	Serial 0/0/1 (S0/0/1)
1900	Gigabit Ethernet 0/0 (G0/0)	Gigabit Ethernet 0/1 (G0/1)	Serial 0/0/0 (S0/0/0)	Serial 0/0/1 (S0/0/1)
2801	Fast Ethernet 0/0 (F0/0)	Fast Ethernet 0/1 (F0/1)	Serial 0/1/0 (S0/1/0)	Serial 0/1/1 (S0/1/1)
2811	Fast Ethernet 0/0 (F0/0)	Fast Ethernet 0/1 (F0/1)	Serial 0/0/0 (S0/0/0)	Serial 0/0/1 (S0/0/1)
2900	Gigabit Ethernet 0/0 (G0/0)	Gigabit Ethernet 0/1 (G0/1)	Serial 0/0/0 (S0/0/0)	Serial 0/0/1 (S0/0/1)
Note: To find out how the router is configured, look at the interfaces to identify the type of router and how many interfaces the router has. There is no way to effectively list all the combinations of configurations for each router class. This table includes identifiers for the possible combinations of Ethernet and Serial interfaces in the device. The table does not include any other type of interface, even though a specific router may contain one. An example of this might be an ISDN BRI interface. The string in parenthesis is the legal abbreviation that can be used in Cisco IOS commands to represent the interface.				

Appendix A: Configuration Commands

Switch S1

```
S1(config)# vlan 10
S1(config-vlan)# name Student
S1(config-vlan)# exit
S1(config)# vlan 20
S1(config-vlan)# name Faculty-Admin
S1(config-vlan)# exit
S1(config)# interface f0/1
S1(config-if)# switchport mode trunk
S1(config-if)# interface range f0/5 - 6
S1(config-if-range)# switchport mode access
S1(config-if-range)# switchport access vlan 10
S1(config-if-range)# interface vlan 10
S1(config-if)# ip address 192.168.10.11 255.255.255.0
S1(config-if)# no shut
S1(config-if)# exit
S1(config)# ip default-gateway 192.168.10.1
```

Switch S2

```
S2(config)# vlan 10
S2(config-vlan)# name Student
S2(config-vlan)# exit
S2(config)# vlan 20
S2(config-vlan)# name Faculty-Admin
S2(config-vlan)# exit
S2(config)# interface f0/1
S2(config-if)# switchport mode trunk
S2(config-if)# interface f0/11
S2(config-if)# switchport mode access
S2(config-if)# switchport access vlan 20
S2(config-if)# interface f0/18
S2(config-if)# switchport mode access
S2(config-if)# switchport access vlan 20
S2(config-if-range)# interface vlan 10
S2(config-if)#ip address 192.168.10.12 255.255.255.0
S2(config-if)# no shut
S2(config-if)# exit
S2(config)# ip default-gateway 192.168.10.1
```

5.1.3.7 Lab – Configuring 802.1Q Trunk-Based Inter-VLAN Routing

Topology

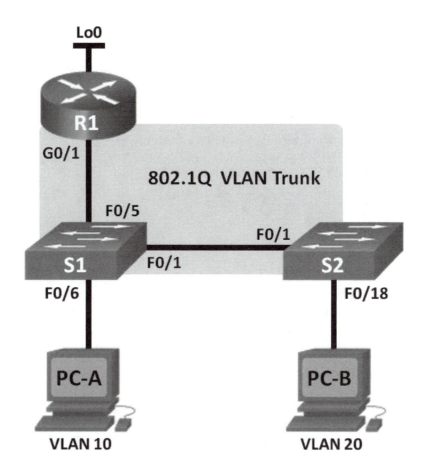

Addressing Table

Device	Interface	IP Address	Subnet Mask	Default Gateway
R1	G0/1.1	192.168.1.1	255.255.255.0	N/A
	G0/1.10	192.168.10.1	255.255.255.0	N/A
	G0/1.20	192.168.20.1	255.255.255.0	N/A
	Lo0	209.165.200.225	255.255.255.224	N/A
S1	VLAN 1	192.168.1.11	255.255.255.0	192.168.1.1
S2	VLAN 1	192.168.1.12	255.255.255.0	192.168.1.1
PC-A	NIC	192.168.10.3	255.255.255.0	192.168.10.1
PC-B	NIC	192.168.20.3	255.255.255.0	192.168.20.1

Switch Port Assignment Specifications

Ports	Assignment	Network
S1 F0/1	802.1Q Trunk	N/A
S2 F0/1	802.1Q Trunk	N/A
S1 F0/5	802.1Q Trunk	N/A
S1 F0/6	VLAN 10 – Students	192.168.10.0/24
S2 F0/18	VLAN 20 – Faculty	192.168.20.0/24

Objectives

Part 1: Build the Network and Configure Basic Device Settings

Part 2: Configure Switches with VLANs and Trunking

Part 3: Configure Trunk-Based Inter-VLAN Routing

Background / Scenario

A second method of providing routing and connectivity for multiple VLANs is through the use of an 802.1Q trunk between one or more switches and a single router interface. This method is also known as router-on-a-stick inter-VLAN routing. In this method, the physical router interface is divided into multiple subinterfaces that provide logical pathways to all VLANs connected.

In this lab, you will configure trunk-based inter-VLAN routing and verify connectivity to hosts on different VLANs as well as with a loopback on the router.

Note: This lab provides minimal assistance with the actual commands necessary to configure trunk-based inter-VLAN routing. However, the required configuration commands are provided in Appendix A of this lab. Test your knowledge by trying to configure the devices without referring to the appendix.

Note: The routers used with CCNA hands-on labs are Cisco 1941 Integrated Services Routers (ISRs) with Cisco IOS, Release 15.2(4)M3 (universalk9 image). The switches used are Cisco Catalyst 2960s with Cisco IOS, Release 15.0(2) (lanbasek9 image). Other routers, switches and Cisco IOS versions can be used. Depending on the model and Cisco IOS version, the commands available and output produced might vary from what is shown in the labs. Refer to the Router Interface Summary Table at the end of the lab for the correct interface identifiers.

Note: Make sure that the routers and switches have been erased and have no startup configurations. If you are unsure, contact your instructor.

Required Resources

- 1 Router (Cisco 1941 with Cisco IOS, release 15.2(4)M3 universal image or comparable)
- 2 Switches (Cisco 2960 with Cisco IOS, release 15.0(2) lanbasek9 image or comparable)
- 2 PCs (Windows 7, Vista, or XP with terminal emulation program, such as Tera Term)
- Console cables to configure the Cisco IOS devices via the console ports
- Ethernet cables as shown in the topology

Part 1:   Build the Network and Configure Basic Device Settings

In Part 1, you will set up the network topology and configure basic settings on the PC hosts, switches, and router.

Step 1:   Cable the network as shown in the topology.

Step 2:   Configure PC hosts.

Step 3:   Initialize and reload the router and switches as necessary.

Step 4:   Configure basic settings for each switch.

    a.   Disable DNS lookup.

    b.   Configure device names as shown in the topology.

    c.   Assign **class** as the privileged EXEC password.

    d.   Assign **cisco** as the console and vty passwords.

    e.   Configure **logging synchronous** for the console line.

    f.   Configure the IP address listed in the Addressing Table for VLAN 1 on both switches.

    g.   Configure the default gateway on both switches.

    h.   Administratively deactivate all unused ports on the switch.

    i.   Copy the running configuration to the startup configuration.

Step 5:   Configure basic settings for the router.

    a.   Disable DNS lookup.

    b.   Configure device names as shown in the topology.

    c.   Configure the Lo0 IP address as shown in the Address Table. Do not configure subinterfaces at this time as they will be configured in Part 3.

    d.   Assign **cisco** as the console and vty passwords.

    e.   Assign **class** as the privileged EXEC password.

    f.   Configure **logging synchronous** to prevent console messages from interrupting command entry.

    g.   Copy the running configuration to the startup configuration.

Part 2: **Configure Switches with VLANs and Trunking**

In Part 2, you will configure the switches with VLANs and trunking.

Note: The required commands for Part 2 are provided in Appendix A. Test your knowledge by trying to configure S1 and S2 without referring to the appendix.

Step 1: **Configure VLANs on S1.**

a. On S1, configure the VLANs and names listed in the Switch Port Assignment Specifications table. Write the commands you used in the space provided.

_____

_____

_____

_____

b. On S1, configure the interface connected to R1 as a trunk. Also configure the interface connected to S2 as a trunk. Write the commands you used in the space provided.

_____

_____

c. On S1, assign the access port for PC-A to VLAN 10. Write the commands you used in the space provided.

_____

_____

_____

Step 2: **Configure VLANs on Switch 2.**

a. On S2, configure the VLANs and names listed in the Switch Port Assignment Specifications table.

b. On S2, verify that the VLAN names and numbers match those on S1. Write the command you used in the space provided.

_____

c. On S2, assign the access port for PC-B to VLAN 20.

d. On S2, configure the interface connected to S1 as a trunk.

Part 3: **Configure Trunk-Based Inter-VLAN Routing**

In Part 3, you will configure R1 to route to multiple VLANs by creating subinterfaces for each VLAN. This method of inter-VLAN routing is called router-on-a-stick.

Note: The required commands for Part 3 are provided in Appendix A. Test your knowledge by trying to configure trunk-based or router-on-a-stick inter-VLAN routing without referring to the appendix.

Step 1: **Configure a subinterface for VLAN 1.**

a. Create a subinterface on R1 G0/1 for VLAN 1 using 1 as the subinterface ID. Write the command you used in the space provided.

_____

b. Configure the subinterface to operate on VLAN 1. Write the command you used in the space provided.

_____

c. Configure the subinterface with the IP address from the Address Table. Write the command you used in the space provided.

_____

Step 2: **Configure a subinterface for VLAN 10.**

a. Create a subinterface on R1 G0/1 for VLAN 10 using 10 as the subinterface ID.

b. Configure the subinterface to operate on VLAN 10.

c. Configure the subinterface with the address from the Address Table.

Step 3: **Configure a subinterface for VLAN 20.**

a. Create a subinterface on R1 G0/1 for VLAN 20 using 20 as the subinterface ID.

b. Configure the subinterface to operate on VLAN 20.

c. Configure the subinterface with the address from the Address Table.

Step 4: **Enable the G0/1 interface.**

Enable the G0/1 interface. Write the commands you used in the space provided.

_____

Step 5:   **Verify connectivity.**

Enter the command to view the routing table on R1. What networks are listed?

_____

From PC-A, is it possible to ping the default gateway for VLAN 10? _____

From PC-A, is it possible to ping PC-B? _____

From PC-A, is it possible to ping Lo0? _____

From PC-A, is it possible to ping S2? _____

If the answer is **no** to any of these questions, troubleshoot the configurations and correct any errors.

Reflection

What are the advantages of trunk-based or router-on-a-stick inter-VLAN routing?

_____

_____

Router Interface Summary Table

Router Interface Summary				
Router Model	**Ethernet Interface #1**	**Ethernet Interface #2**	**Serial Interface #1**	**Serial Interface #2**
1800	Fast Ethernet 0/0 (F0/0)	Fast Ethernet 0/1 (F0/1)	Serial 0/0/0 (S0/0/0)	Serial 0/0/1 (S0/0/1)
1900	Gigabit Ethernet 0/0 (G0/0)	Gigabit Ethernet 0/1 (G0/1)	Serial 0/0/0 (S0/0/0)	Serial 0/0/1 (S0/0/1)
2801	Fast Ethernet 0/0 (F0/0)	Fast Ethernet 0/1 (F0/1)	Serial 0/1/0 (S0/1/0)	Serial 0/1/1 (S0/1/1)
2811	Fast Ethernet 0/0 (F0/0)	Fast Ethernet 0/1 (F0/1)	Serial 0/0/0 (S0/0/0)	Serial 0/0/1 (S0/0/1)
2900	Gigabit Ethernet 0/0 (G0/0)	Gigabit Ethernet 0/1 (G0/1)	Serial 0/0/0 (S0/0/0)	Serial 0/0/1 (S0/0/1)

Note: To find out how the router is configured, look at the interfaces to identify the type of router and how many interfaces the router has. There is no way to effectively list all the combinations of configurations for each router class. This table includes identifiers for the possible combinations of Ethernet and Serial interfaces in the device. The table does not include any other type of interface, even though a specific router may contain one. An example of this might be an ISDN BRI interface. The string in parenthesis is the legal abbreviation that can be used in Cisco IOS commands to represent the interface.

Appendix A – Configuration Commands

Switch S1

```
S1(config)# vlan 10
S1(config-vlan)# name Students
S1(config-vlan)# vlan 20
S1(config-vlan)# name Faculty
S1(config-vlan)# exit
S1(config)# interface f0/1
S1(config-if)# switchport mode trunk
S1(config-if)# interface f0/5
S1(config-if)# switchport mode trunk
S1(config-if)# interface f0/6
S1(config-if)# switchport mode access
S1(config-if)# switchport access vlan 10
```

Switch S2

```
S2(config)# vlan 10
S2(config-vlan)# name Students
S2(config-vlan)# vlan 20
S2(config-vlan)# name Faculty
S2(config)# interface f0/1
S2(config-if)# switchport mode trunk
S2(config-if)# interface f0/18
S2(config-if)# switchport mode access
S2(config-if)# switchport access vlan 20
```

Router R1

```
R1(config)# interface g0/1.1
R1(config-subif)# encapsulation dot1Q 1
R1(config-subif)# ip address 192.168.1.1 255.255.255.0
R1(config-subif)# interface g0/1.10
R1(config-subif)# encapsulation dot1Q 10
R1(config-subif)# ip address 192.168.10.1 255.255.255.0
R1(config-subif)# interface g0/1.20
R1(config-subif)# encapsulation dot1Q 20
R1(config-subif)# ip address 192.168.20.1 255.255.255.0
R1(config-subif)# exit
R1(config)# interface g0/1
R1(config-if)# no shutdown
```

5.3.2.4 Lab – Troubleshooting Inter-VLAN Routing

Topology

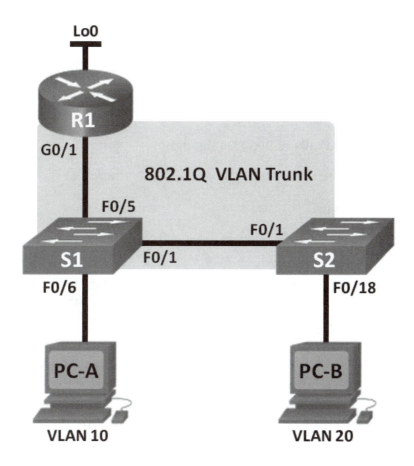

Addressing Table

Device	Interface	IP Address	Subnet Mask	Default Gateway
R1	G0/1.1	192.168.1.1	255.255.255.0	N/A
	G0/1.10	192.168.10.1	255.255.255.0	N/A
	G0/1.20	192.168.20.1	255.255.255.0	N/A
	Lo0	209.165.200.225	255.255.255.224	N/A
S1	VLAN 1	192.168.1.11	255.255.255.0	192.168.1.1
S2	VLAN 1	192.168.1.12	255.255.255.0	192.168.1.1
PC-A	NIC	192.168.10.3	255.255.255.0	192.168.10.1
PC-B	NIC	192.168.20.3	255.255.255.0	192.168.20.1

Switch Port Assignment Specifications

Ports	Assignment	Network
S1 F0/1	802.1Q Trunk	N/A
S2 F0/1	802.1Q Trunk	N/A
S1 F0/5	802.1Q Trunk	N/A
S1 F0/6	VLAN 10 – R&D	192.168.10.0/24
S2 F0/18	VLAN 20 – Engineering	192.168.20.0/24

Objectives

Part 1: Build the Network and Load Device Configurations

Part 2: Troubleshoot the Inter-VLAN Routing Configuration

Part 3: Verify VLAN Configuration, Port Assignment, and Trunking

Part 4: Test Layer 3 Connectivity

Background / Scenario

The network has been designed and configured to support three VLANs. Inter-VLAN routing is provided by an external router using an 802.1Q trunk, also known as router-on-a-stick. Routing to a remote web server, which is simulated by Lo0, is also provided by R1. However, it is not working as designed, and user complaints have not given much insight into the source of the problems.

In this lab, you must first define what is not working as expected, and then analyze the existing configurations to determine and correct the source of the problems. This lab is complete when you can demonstrate IP connectivity between each of the user VLANs and the external web server network, and between the switch management VLAN and the web server network.

Note: The routers used with CCNA hands-on labs are Cisco 1941 Integrated Services Routers (ISRs) with Cisco IOS Release 15.2(4)M3 (universalk9 image). The switches used are Cisco Catalyst 2960s with Cisco IOS Release 15.0(2) (lanbasek9 image). Other routers, switches, and Cisco IOS versions can be used. Depending on the model and Cisco IOS version, the commands available and output produced might vary from what is shown in the labs. Refer to the Router Interface Summary Table at the end of this lab for the correct interface identifiers.

Note: Make sure that the routers and switches have been erased and have no startup configurations. If you are unsure, contact your instructor.

Required Resources

- 1 Router (Cisco 1941 with Cisco IOS Release 15.2(4)M3 universal image or comparable)
- 2 Switches (Cisco 2960 with Cisco IOS Release 15.0(2) lanbasek9 image or comparable)
- 2 PCs (Windows 7, Vista, or XP with terminal emulation program, such as Tera Term)
- Console cables to configure the Cisco IOS devices via the console ports
- Ethernet cables as shown in the topology

Part 1: Build the Network and Load Device Configurations

In Part 1, you will set up the network topology and configure basic settings on the PC hosts, switches, and router.

Step 1:   Cable the network as shown in the topology.

Step 2:   Configure PC hosts.

Refer to the Addressing Table for PC host address information.

Step 3:   Load router and switch configurations.

Load the following configurations into the appropriate router or switch. All devices have the same passwords; the enable password is **class**, and the line password is **cisco**.

Router R1 Configuration:

```
hostname R1
enable secret class
no ip domain lookup
line con 0
 password cisco
 login
 logging synchronous
line vty 0 4
 password cisco
 login
interface loopback0
 ip address 209.165.200.225 255.255.255.224
interface gigabitEthernet0/1
 no ip address

interface gigabitEthernet0/1.1
 encapsulation dot1q 11

 ip address 192.168.1.1 255.255.255.0
interface gigabitEthernet0/1.10
 encapsulation dot1q 10
 ip address 192.168.11.1 255.255.255.0

interface gigabitEthernet0/1.20
 encapsulation dot1q 20
 ip address 192.168.20.1 255.255.255.0
end
```

Switch S1 Configuration:

```
hostname S1
enable secret class
no ip domain-lookup
line con 0
 password cisco
 login
 logging synchronous
line vty 0 15
 password cisco
 login
vlan 10
 name R&D
 exit




interface fastethernet0/1
 switchport mode access

interface fastethernet0/5
 switchport mode trunk



interface vlan1
  ip address 192.168.1.11 255.255.255.0
ip default-gateway 192.168.1.1
end
```

Switch S2 Configuration:

```
hostname S2
enable secret class
no ip domain-lookup
line con 0
 password cisco
 login
 logging synchronous
line vty 0 15
```

```
 password cisco
 login



vlan 20
 name Engineering
 exit
interface fastethernet0/1
 switchport mode trunk
interface fastethernet0/18
 switchport access vlan 10
 switchport mode access

interface vlan1
 ip address 192.168.1.12 255.255.255.0
ip default-gateway 192.168.1.1
end
```

Step 4:   **Save the running configuration to the startup configuration.**

Part 2:   **Troubleshoot the Inter-VLAN Routing Configuration**

In Part 2, you will verify the inter-VLAN routing configuration.

a.   On R1, enter the **show ip route** command to view the routing table.

Which networks are listed?

_____

Are there any networks missing in the routing table? If so, which networks?

_____

What is one possible reason that a route would be missing from the routing table?

_____

b. On R1, issue the **show ip interface brief** command.

Based on the output, are there any interface issues on the router? If so, what commands would resolve the issues?

_____

_____

_____

c. On R1, re-issue the **show ip route** command.

Verify that all networks are available in the routing table. If not, continue to troubleshoot until all networks are present.

Part 3: Verify VLAN Configuration, Port Assignment, and Trunking

In Part 3, you will verify that the correct VLANs exist on both S1 and S2 and that trunking is configured correctly.

Step 1: Verify VLAN configuration and port assignments.

a. On S1, enter the **show vlan brief** command to view the VLAN database.

Which VLANs are listed? Ignore VLANs 1002 to 1005.

_____

Are there any VLANs numbers or names missing in the output? If so, list them.

_____

Are the access ports assigned to the correct VLANs? If not, list the missing or incorrect assignments.

_____

If required, what commands would resolve the VLAN issues?

_____

_____

_____

b. On S1, re-issue the **show vlan brief** command to verify configuration.

c. On S2, enter the **show vlan brief** command to view the VLAN database.

Which VLANs are listed? Ignore VLANs 1002 to 1005.

_____

Are there any VLANs numbers or names missing in the output? If so, list them.

_____

Are the access ports assigned to the correct VLANs? If not, list the missing or incorrect assignments.

_____

If required, what commands would resolve the VLAN issues?

_____

_____

_____

d. On S2, re-issue the **show vlan brief** command to verify any configuration changes.

Step 2: **Verify trunking interfaces.**

a. On S1, enter the **show interface trunk** command to view the trunking interfaces.

Which ports are in trunking mode?

_____

Are there any ports missing in the output? If so, list them.

_____

If required, what commands would resolve the port trunking issues?

_____

_____

b. On S1, re-issue the **show interface trunk** command to verify any configuration changes.

c. On S2, enter the **show interface trunk** command to view the trunking interfaces.

Which ports are in trunking mode?

_____

Are there any ports missing in the output? If so, list them.

_____

If required, what commands would resolve the port trunking issues?

_____

_____

Part 4: Test Layer 3 Connectivity

a. Now that you have corrected multiple configuration issues, let's test connectivity.

From PC-A, is it possible to ping the default gateway for VLAN 10? _____

From PC-A, is it possible to ping PC-B? _____

From PC-A, is it possible to ping Lo0? _____

If the answer is **no** to any of these questions, troubleshoot the configurations and correct the error.

Note: It may be necessary to disable the PC firewall for pings between PCs to be successful.

From PC-A, is it possible to ping S1? _____

From PC-A, is it possible to ping S2? _____

List some of the issues that could still be preventing successful pings to the switches.

_____

_____

b. One way to help resolve where the error is occurring is to do a **tracert** from PC-A to S1.

```
C:\Users\User1> tracert 192.168.1.11

Tracing route to 192.168.1.11 over a maximum of 30 hops

   1    <1 ms    <1 ms    <1 ms   192.168.10.1
   2     *        *        *       Request timed out.
   3     *        *        *       Request timed out.
<output omitted>
```

This output shows that the request from PC-A is reaching the default gateway on R1 g0/1.10, but the packet stops at the router.

c. You have already verified the routing table entries for R1, now execute the **show run | section interface** command to verify VLAN configuration. List any configuration errors.

_____

_____

What commands would resolve any issues found?

_____

_____

_____

d. Verify that that pings from PC-A now reach both S1 and S2.

From PC-A, is it possible to ping S1? _____

From PC-A, is it possible to ping S2? _____

Reflection

What are the advantages of viewing the routing table for troubleshooting purposes?

_____

_____

Router Interface Summary Table

Router Interface Summary				
Router Model	**Ethernet Interface #1**	**Ethernet Interface #2**	**Serial Interface #1**	**Serial Interface #2**
1800	Fast Ethernet 0/0 (F0/0)	Fast Ethernet 0/1 (F0/1)	Serial 0/0/0 (S0/0/0)	Serial 0/0/1 (S0/0/1)
1900	Gigabit Ethernet 0/0 (G0/0)	Gigabit Ethernet 0/1 (G0/1)	Serial 0/0/0 (S0/0/0)	Serial 0/0/1 (S0/0/1)
2801	Fast Ethernet 0/0 (F0/0)	Fast Ethernet 0/1 (F0/1)	Serial 0/1/0 (S0/1/0)	Serial 0/1/1 (S0/1/1)
2811	Fast Ethernet 0/0 (F0/0)	Fast Ethernet 0/1 (F0/1)	Serial 0/0/0 (S0/0/0)	Serial 0/0/1 (S0/0/1)
2900	Gigabit Ethernet 0/0 (G0/0)	Gigabit Ethernet 0/1 (G0/1)	Serial 0/0/0 (S0/0/0)	Serial 0/0/1 (S0/0/1)
Note: To find out how the router is configured, look at the interfaces to identify the type of router and how many interfaces the router has. There is no way to effectively list all the combinations of configurations for each router class. This table includes identifiers for the possible combinations of Ethernet and Serial interfaces in the device. The table does not include any other type of interface, even though a specific router may contain one. An example of this might be an ISDN BRI interface. The string in parenthesis is the legal abbreviation that can be used in Cisco IOS commands to represent the interface.				

5.4.1.1 Class Activity – The Inside Track

Objective

Explain how Layer 3 switches forward data in a small- to medium-sized business LAN.

Scenario

Your company has just purchased a three-level building. You are the network administrator and must design the company inter-VLAN routing network scheme to serve a few employees on each floor.

Floor 1 is occupied by the HR Department, Floor 2 is occupied by the IT Department, and Floor 3 is occupied by the Sales Department. All Departments must be able to communicate with each other, but at the same time have their own separate working networks.

You brought three Cisco 2960 switches and a Cisco 1941 series router from the old office location to serve network connectivity in the new building. New equipment is non-negotiable.

Refer to the PDF for this activity for further instructions.

Resources

- Software presentation program

Directions

Work with a partner to complete this activity.

Step 1:   **Design your topology.**

    a.   Use one 2960 switch per floor of your new building.

    b.   Assign one department to each switch.

    c.   Pick one of the switches to connect to the 1941 series router.

Step 2:   **Plan the VLAN scheme.**

    a.   Devise VLAN names and numbers for the HR, IT, and Sales Departments.

    b.   Include a management VLAN, possibly named Management or Native, numbered to your choosing.

    c.   Use either IPv4 or v6 as your addressing scheme for the LANs. If using IPv4, you must also use VLSM.

Step 3:   **Design a graphic to show your VLAN design and address scheme.**

Step 4:   **Choose your inter-VLAN routing method.**

    a.   Legacy (per interface)

    b.   Router-on-a-Stick

    c.   Multilayer switching

Step 5: Create a presentation justifying your inter-VLAN routing method of choice.

a. No more than eight slides can be created for the presentation.

b. Present your group's design to the class or to your instructor.

    1) Be able to explain the method you chose. What makes it different or more desirable to your business than the other two methods?

    2) Be able to show how data moves throughout your network. Verbally explain how the networks are able to communicate using your inter-VLAN method of choice.

Chapter 6 — Static Routing

6.0.1.2 Class Activity – Which Way Should We Go?

Objectives

Explain the benefits of using static routes.

Scenario

A huge sporting event is about to take place in your city. To attend the event, you make concise plans to arrive at the sports arena on time to see the entire game.

There are two routes you can take to drive to the event:

- Highway route - It is easy to follow and fast driving speeds are allowed.

- Alternative, direct route - You found this route using a city map. Depending on conditions, such as the amount of traffic or congestion, this just may be the way to get to the arena on time!

With a partner, discuss these options. Choose a preferred route to arrive at the arena in time to see every second of the huge sporting event.

Compare your optional preferences to network traffic, which route would you choose to deliver data communications for your small- to medium-sized business? Would it be the fastest, easiest route or the alternative, direct route? Justify your choice.

Complete the modeling activity .pdf and be prepared to justify your answers to the class or with another group.

Required Resources

None

Reflection

1. Which route did you choose as your first preference? On what criteria did you base your decision?

_____

2. If traffic congestion were to occur on either route, would this change the path you would take to the arena? Explain your answer.

_____

3.   A popular phrase that can be argued is "the shortest distance between two points is a straight line." Is this always true with delivery of network data? How do you compare your answer to this modeling activity scenario?

_____

6.2.2.5 Lab – Configuring IPv4 Static and Default Routes

Topology

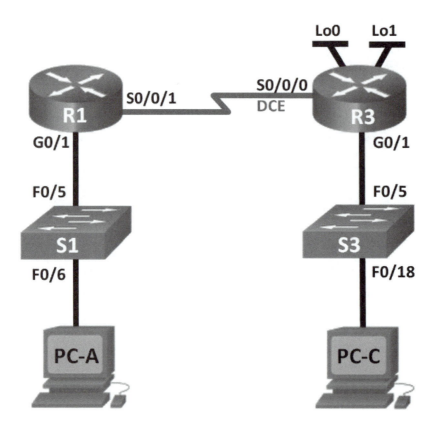

Addressing Table

Device	Interface	IP Address	Subnet Mask	Default Gateway
R1	G0/1	192.168.0.1	255.255.255.0	N/A
	S0/0/1	10.1.1.1	255.255.255.252	N/A
R3	G0/1	192.168.1.1	255.255.255.0	N/A
	S0/0/0 (DCE)	10.1.1.2	255.255.255.252	N/A
	Lo0	209.165.200.225	255.255.255.224	N/A
	Lo1	198.133.219.1	255.255.255.0	N/A
PC-A	NIC	192.168.0.10	255.255.255.0	192.168.0.1
PC-C	NIC	192.168.1.10	255.255.255.0	192.168.1.1

Objectives

Part 1: Set Up the Topology and Initialize Devices

Part 2: Configure Basic Device Settings and Verify Connectivity

Part 3: Configure Static Routes

- Configure a recursive static route.

- Configure a directly connected static route.

- Configure and remove static routes.

Part 4: Configure and Verify a Default Route

Background / Scenario

A router uses a routing table to determine where to send packets. The routing table contains a set of routes that describe which gateway or interface the router uses to reach a specified network. Initially, the routing table contains only directly connected networks. To communicate with distant networks, routes must be specified and added to the routing table.

In this lab, you will manually configure a static route to a specified distant network based on a next-hop IP address or exit interface. You will also configure a static default route. A default route is a type of static route that specifies a gateway to use when the routing table does not contain a path for the destination network.

Note: This lab provides minimal assistance with the actual commands necessary to configure static routing. However, the required commands are provided in Appendix A. Test your knowledge by trying to configure the devices without referring to the appendix.

Note: The routers used with CCNA hands-on labs are Cisco 1941 Integrated Services Routers (ISRs) with Cisco IOS Release 15.2(4)M3 (universalk9 image). The switches used are Cisco Catalyst 2960s with Cisco IOS Release 15.0(2) (lanbasek9 image). Other routers, switches, and Cisco IOS versions can be used. Depending on the model and Cisco IOS version, the commands available and output produced might vary from what is shown in the labs. Refer to the Router Interface Summary Table at the end of this lab for the correct interface identifiers.

Note: Make sure that the routers and switches have been erased and have no startup configurations. If you are unsure, contact your instructor.

Required Resources

- 2 Routers (Cisco 1941 with Cisco IOS Release 15.2(4)M3 universal image or comparable)

- 2 Switches (Cisco 2960 with Cisco IOS Release 15.0(2) lanbasek9 image or comparable)

- 2 PCs (Windows 7, Vista, or XP with terminal emulation program, such as Tera Term)

- Console cables to configure the Cisco IOS devices via the console ports

- Ethernet and serial cables as shown in the topology

Part 1:   Set Up the Topology and Initialize Devices

Step 1:   **Cable the network as shown in the topology.**

Step 2:   **Initialize and reload the router and switch.**

Part 2:   Configure Basic Device Settings and Verify Connectivity

In Part 2, you will configure basic settings, such as the interface IP addresses, device access, and passwords. You will verify LAN connectivity and identify routes listed in the routing tables for R1 and R3.

Step 1:   **Configure the PC interfaces.**

Step 2:   **Configure basic settings on the routers.**

   a.   Configure device names, as shown in the Topology and Addressing Table.

   b.   Disable DNS lookup.

   c.   Assign **class** as the enable password and assign **cisco** as the console and vty password.

   d.   Save the running configuration to the startup configuration file.

Step 3:   **Configure IP settings on the routers.**

   a.   Configure the R1 and R3 interfaces with IP addresses according to the Addressing Table.

   b.   The S0/0/0 connection is the DCE connection and requires the **clock rate** command. The R3 S0/0/0 configuration is displayed below.

```
R3(config)# interface s0/0/0
R3(config-if)# ip address 10.1.1.2 255.255.255.252
R3(config-if)# clock rate 128000
R3(config-if)# no shutdown
```

Step 4:   **Verify connectivity of the LANs.**

   a.   Test connectivity by pinging from each PC to the default gateway that has been configured for that host.

      From PC-A, is it possible to ping the default gateway? _____

      From PC-C, is it possible to ping the default gateway? _____

   b.   Test connectivity by pinging between the directly connected routers.

      From R1, is it possible to ping the S0/0/0 interface of R3? _____

      If the answer is **no** to any of these questions, troubleshoot the configurations and correct the error.

   c.   Test connectivity between devices that are not directly connected.

      From PC-A, is it possible to ping PC-C? _____

      From PC-A, is it possible to ping Lo0? _____

      From PC-A, is it possible to ping Lo1? _____

      Were these pings successful? Why or why not?

      _____

      _____

   Note: It may be necessary to disable the PC firewall to ping between PCs.

Step 5:   **Gather information.**

a.   Check the status of the interfaces on R1 with the **show ip interface brief** command.

How many interfaces are activated on R1? _____

b.   Check the status of the interfaces on R3.

How many interfaces are activated on R3? _____

c.   View the routing table information for R1 using the **show ip route** command.

What networks are present in the Addressing Table of this lab, but not in the routing table for R1?

_____

d. View the routing table information for R3.

What networks are present in the Addressing Table in this lab, but not in the routing table for R3?

_____

Why are all the networks not in the routing tables for each of the routers?

_____

_____

Part 3:   **Configure Static Routes**

In Part 3, you will employ multiple ways to implement static and default routes, you will confirm that the routes have been added to the routing tables of R1 and R3, and you will verify connectivity based on the introduced routes.

Note: This lab provides minimal assistance with the actual commands necessary to configure static routing. However, the required commands are provided in Appendix A. Test your knowledge by trying to configure the devices without referring to the appendix.

Step 1:   **Configure a recursive static route.**

With a recursive static route, the next-hop IP address is specified. Because only the next-hop IP is specified, the router must perform multiple lookups in the routing table before forwarding packets. To configure recursive static routes, use the following syntax:

```
Router(config)# ip route network-address subnet-mask ip-address
```

a.   On the R1 router, configure a static route to the 192.168.1.0 network using the IP address of the Serial 0/0/0 interface of R3 as the next-hop address. Write the command you used in the space provided.

_____

b.   View the routing table to verify the new static route entry.

How is this new route listed in the routing table?

_____

From host PC-A, is it possible to ping the host PC-C? _____

These pings should fail. If the recursive static route is correctly configured, the ping arrives at PC-C. PC-C sends a ping reply back to PC-A. However, the ping reply is discarded at R3 because R3 does not have a return route to the 192.168.0.0 network in the routing table.

Step 2: Configure a directly connected static route.

With a directly connected static route, the *exit-interface* parameter is specified, which allows the router to resolve a forwarding decision in one lookup. A directly connected static route is typically used with a point-to-point serial interface. To configure directly connected static routes with an exit interface specified, use the following syntax:

```
Router(config)# ip route network-address subnet-mask exit-intf
```

a. On the R3 router, configure a static route to the 192.168.0.0 network using S0/0/0 as the exit interface. Write the command you used in the space provided.

_____

b. View the routing table to verify the new static route entry.

How is this new route listed in the routing table?

_____

c. From host PC-A, is it possible to ping the host PC-C? _____

This ping should be successful.

Note: It may be necessary to disable the PC firewall to ping between PCs.

Step 3: **Configure a static route.**

a. On the R1 router, configure a static route to the 198.133.219.0 network using one of the static route configuration options from the previous steps. Write the command you used in the space provided.

_____

b. On the R1 router, configure a static route to the 209.165.200.224 network on R3 using the other static route configuration option from the previous steps. Write the command you used in the space provided.

_____

c. View the routing table to verify the new static route entry.

How is this new route listed in the routing table?

_____

d. From host PC-A, is it possible to ping the R1 address 198.133.219.1? _____

This ping should be successful.

Step 4:   Remove static routes for loopback addresses.

a. On R1, use the **no** command to remove the static routes for the two loopback addresses from the routing table. Write the commands you used in the space provided.

_____

b. View the routing table to verify the routes have been removed.

How many network routes are listed in the routing table on R1? _____

Is the Gateway of last resort set? _____

Part 4:   Configure and Verify a Default Route

In Part 4, you will implement a default route, confirm that the route has been added to the routing table, and verify connectivity based on the introduced route.

A default route identifies the gateway to which the router sends all IP packets for which it does not have a learned or static route. A default static route is a static route with 0.0.0.0 as the destination IP address and subnet mask. This is commonly referred to as a "quad zero" route.

In a default route, either the next-hop IP address or exit interface can be specified. To configure a default static route, use the following syntax:

```
Router(config)# ip route 0.0.0.0 0.0.0.0 {ip-address or exit-intf}
```

a.   Configure the R1 router with a default route using the exit interface of S0/0/1. Write the command you used in the space provided.

_____

b.   View the routing table to verify the new static route entry.

How is this new route listed in the routing table?

_____

What is the Gateway of last resort?

_____

c.   From host PC-A, is it possible to ping the 209.165.200.225? _____

d.   From host PC-A, is it possible to ping the 198.133.219.1? _____

These pings should be successful.

Reflection

1. A new network 192.168.3.0/24 is connected to interface G0/0 on R1. What commands could be used to configure a static route to that network from R3?

_____

_____

2. Is there a benefit to configuring a directly connected static route instead of a recursive static route?

_____

_____

3. Why is it important to configure a default route on a router?

_____

Router Interface Summary Table

Router Interface Summary				
Router Model	**Ethernet Interface #1**	**Ethernet Interface #2**	**Serial Interface #1**	**Serial Interface #2**
1800	Fast Ethernet 0/0 (F0/0)	Fast Ethernet 0/1 (F0/1)	Serial 0/0/0 (S0/0/0)	Serial 0/0/1 (S0/0/1)
1900	Gigabit Ethernet 0/0 (G0/0)	Gigabit Ethernet 0/1 (G0/1)	Serial 0/0/0 (S0/0/0)	Serial 0/0/1 (S0/0/1)
2801	Fast Ethernet 0/0 (F0/0)	Fast Ethernet 0/1 (F0/1)	Serial 0/1/0 (S0/1/0)	Serial 0/1/1 (S0/1/1)
2811	Fast Ethernet 0/0 (F0/0)	Fast Ethernet 0/1 (F0/1)	Serial 0/0/0 (S0/0/0)	Serial 0/0/1 (S0/0/1)
2900	Gigabit Ethernet 0/0 (G0/0)	Gigabit Ethernet 0/1 (G0/1)	Serial 0/0/0 (S0/0/0)	Serial 0/0/1 (S0/0/1)
Note: To find out how the router is configured, look at the interfaces to identify the type of router and how many interfaces the router has. There is no way to effectively list all the combinations of configurations for each router class. This table includes identifiers for the possible combinations of Ethernet and Serial interfaces in the device. The table does not include any other type of interface, even though a specific router may contain one. An example of this might be an ISDN BRI interface. The string in parenthesis is the legal abbreviation that can be used in Cisco IOS commands to represent the interface.				

Appendix A: Configuration Commands for Parts 2, 3, and 4

The commands listed in Appendix A are for reference only. This Appendix does not include all the specific commands necessary to complete this lab.

Basic Device Settings

Configure IP settings on the router.

```
R3(config)# interface s0/0/0
R3(config-if)# ip address 10.1.1.2 255.255.255.252
R3(config-if)# clock rate 128000
R3(config-if)# no shutdown
```

Static Route Configurations

Configure a recursive static route.

```
R1(config)# ip route 192.168.1.0 255.255.255.0 10.1.1.2
```

Configure a directly connected static route.

```
R3(config)# ip route 192.168.0.0 255.255.255.0 s0/0/0
```

Remove static routes.

```
R1(config)# no ip route 209.165.200.224 255.255.255.224 serial0/0/1
```

or

```
R1(config)# no ip route 209.165.200.224 255.255.255.224 10.1.1.2
```

or

```
R1(config)# no ip route 209.165.200.224 255.255.255.224
```

Default Route Configuration

```
R1(config)# ip route 0.0.0.0 0.0.0.0 s0/0/1
```

6.2.4.5 Lab – Configuring IPv6 Static and Default Routes

Topology

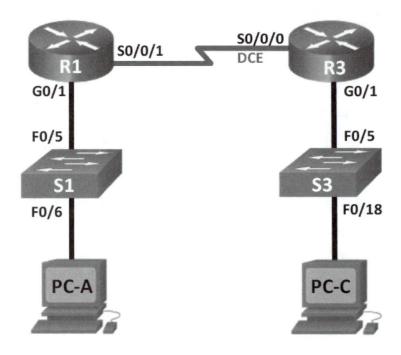

Addressing Table

Device	Interface	IPv6 Address / Prefix Length	Default Gateway
R1	G0/1	2001:DB8:ACAD:A::/64 eui-64	N/A
	S0/0/1	FC00::1/64	N/A
R3	G0/1	2001:DB8:ACAD:B::/64 eui-64	N/A
	S0/0/0	FC00::2/64	N/A
PC-A	NIC	SLAAC	SLAAC
PC-C	NIC	SLAAC	SLAAC

Objectives

Part 1: Build the Network and Configure Basic Device Settings

- Enable IPv6 unicast routing and configure IPv6 addressing on the routers.
- Disable IPv4 addressing and enable IPv6 SLAAC for the PC network interfaces.
- Use **ipconfig** and **ping** to verify LAN connectivity.
- Use **show** commands to verify IPv6 settings.

Part 2: Configure IPv6 Static and Default Routes

- Configure a directly attached IPv6 static route.
- Configure a recursive IPv6 static route.
- Configure a default IPv6 static route.

Background / Scenario

In this lab, you will configure the entire network to communicate using only IPv6 addressing, including configuring the routers and PCs. You will use stateless address auto-configuration (SLAAC) for configuring the IPv6 addresses for the hosts. You will also configure IPv6 static and default routes on the routers to enable communication to remote networks that are not directly connected.

Note: The routers used with CCNA hands-on labs are Cisco 1941 Integrated Services Routers (ISRs) with Cisco IOS Release 15.2(4)M3 (universalk9 image). The switches used are Cisco Catalyst 2960s with Cisco IOS Release 15.0(2) (lanbasek9 image). Other routers, switches, and Cisco IOS versions can be used. Depending on the model and Cisco IOS version, the commands available and output produced might vary from what is shown in the labs. Refer to the Router Interface Summary Table at the end of this lab for the correct interface identifiers.

Note: Make sure that the routers and switches have been erased and have no startup configurations. If you are unsure, contact your instructor.

Required Resources

- 2 Routers (Cisco 1941 with Cisco IOS Release 15.2(4)M3 universal image or comparable)
- 2 Switches (Cisco 2960 with Cisco IOS Release 15.0(2) lanbasek9 image or comparable)
- 2 PCs (Windows 7, Vista, or XP with terminal emulation program, such as Tera Term)
- Console cables to configure the Cisco IOS devices via the console ports
- Ethernet and serial cables as shown in the topology

Part 1: Build the Network and Configure Basic Device Settings

In Part 1, you will cable and configure the network to communicate using IPv6 addressing.

Step 1: **Cable the network as shown in the topology diagram.**

Step 2: **Initialize and reload the routers and switches.**

Step 3: **Enable IPv6 unicast routing and configure IPv6 addressing on the routers.**

a. Using Tera Term, console into the router labeled R1 in the topology diagram and assign the router the name R1.

b. Within global configuration mode, enable IPv6 routing on R1.

```
R1(config)# ipv6 unicast-routing
```

c. Configure the network interfaces on R1 with IPv6 addresses. Notice that IPv6 is enabled on each interface. The G0/1 interface has a globally routable unicast address and EUI-64 is used to create the interface identifier portion of the address. The S0/0/1 interface has a privately routable, unique-local address, which is recommended for point-to-point serial connections.

```
R1(config)# interface g0/1
R1(config-if)# ipv6 address 2001:DB8:ACAD:A::/64 eui-64
R1(config-if)# no shutdown
R1(config-if)# interface serial 0/0/1
R1(config-if)# ipv6 address FC00::1/64
R1(config-if)# no shutdown
R1(config-if)# exit
```

d. Assign a device name to router R3.

e. Within global configuration mode, enable IPv6 routing on R3.

```
R3(config)# ipv6 unicast-routing
```

f. Configure the network interfaces on R3 with IPv6 addresses. Notice that IPv6 is enabled on each interface. The G0/1 interface has a globally routable unicast address and EUI-64 is used to create the interface identifier portion of the address. The S0/0/0 interface has a privately routable, unique-local address, which is recommended for point-to-point serial connections. The clock rate is set because it is the DCE end of the serial cable.

```
R3(config)# interface gigabit 0/1

R3(config-if)# ipv6 address 2001:DB8:ACAD:B::/64 eui-64

R3(config-if)# no shutdown

R3(config-if)# interface serial 0/0/0

R3(config-if)# ipv6 address FC00::2/64

R3(config-if)# clock rate 128000

R3(config-if)# no shutdown

R3(config-if)# exit
```

Step 4: Disable IPv4 addressing and enable IPv6 SLAAC for the PC network interfaces.

a. On both PC-A and PC-C, navigate to the **Start** menu > **Control Panel**. Click the **Network and Sharing Center** link while viewing with icons. In the Network and Sharing Center window, click the **Change adapter settings** link on the left side of the window to open the Network Connections window.

b. In the Network Connections window, you see the icons for your network interface adapters. Double-click the Local Area Connection icon for the PC network interface that is connected to the switch. Click the **Properties** to open the Local Area Connection Properties dialogue window.

c. With the Local Area Connection Properties window open, scroll down through the items and uncheck the item **Internet Protocol Version 4 (TCP/IPv4)** check box to disable the IPv4 protocol on the network interface.

d. With the Local Area Connection Properties window still open, click the **Internet Protocol Version 6 (TCP/IPv6)** check box, and then click **Properties**.

e. With the Internet Protocol Version 6 (TCP/IPv6) Properties window open, check to see if the radio buttons for **Obtain an IPv6 address automatically** and **Obtain DNS server address automatically** are selected. If not, select them.

f. With the PCs configured to obtain an IPv6 address automatically, they will contact the routers to obtain the network subnet and gateway information, and auto-configure their IPv6 address information. In the next step, you will verify the settings.

Step 5: Use ipconfig and ping to verify LAN connectivity.

a. From PC-A, open a command prompt, type **ipconfig /all** and press Enter. The output should look similar to that shown below. In the output, you should see that the PC now has an IPv6 global unicast address, a link-local IPv6 address, and a link-local IPv6 default gateway address. You may also see a temporary IPv6 address and under the DNS server addresses, three site-local addresses that start with FEC0. Site-local addresses are private addresses that were meant to be backwards compatible with NAT. However, they are not supported in IPv6 and are replaced by unique-local addresses.

```
C:\Users\User1> ipconfig /all

Windows IP Configuration


<Output omitted>


Ethernet adapter Local Area Connection:

    Connection-specific DNS Suffix. . . :
    Description . . . . . . . . . . . : Intel(R) 82577LC Gigabit Network Connection
    Physical Address. . . . . . . . . : 1C-C1-DE-91-C3-5D
    DHCP Enabled. . . . . . . . . . . : No
    Autoconfiguration Enabled. . . . . : Yes
    IPv6 Address. . . . . . . . . . . : 2001:db8:acad:a:7c0c:7493:218d:2f6c(Preferred)
    Temporary IPv6 Address. . . . . . : 2001:db8:acad:a:bc40:133a:54e7:d497(Preferred)
    Link-local IPv6 Address . . . . . : fe80::7c0c:7493:218d:2f6c%13(Preferred)
    Default Gateway . . . . . . . . . : fe80::6273:5cff:fe0d:1a61%13
    DNS Servers . . . . . . . . . . . : fec0:0:0:ffff::1%1
                                        fec0:0:0:ffff::2%1
                                        fec0:0:0:ffff::3%1
    NetBIOS over Tcpip. . . . . . . . : Disabled
```

Based on your network implementation and the output of the **ipconfig /all** command, did PC-A receive IPv6 addressing information from R1?

_____

_____

_____

b. What is the PC-A global unicast IPv6 address?

_____

c. What is the PC-A link-local IPv6 address?

_____

d. What is the PC-A default gateway IPv6 address?

_____

e. From PC-A, use the **ping -6** command to issue an IPv6 ping to the link-local default gateway address. You should see replies from the R1 router.

```
C:\Users\User1> ping -6 <default-gateway-address>
```

Did PC-A receive replies to the ping from PC-A to R1?

_____

f. Repeat Step 5a from PC-C.

Did PC-C receive IPv6 addressing information from R3?

_____

g. What is the PC-C global unicast IPv6 address?

_____

h. What is the PC-C link-local IPv6 address?

_____

i. What is the PC-C default gateway IPv6 address?

_____

j. From PC-C, use the **ping -6** command to ping the PC-C default gateway.

Did PC-C receive replies to the pings from PC-C to R3?

_____

k. Attempt an IPv6 **ping -6** from PC-A to the PC-C IPv6 address.

`C:\Users\User1> `**`ping -6`**` `*`PC-C-IPv6-address`*

Was the ping successful? Why or why not?

_____

_____

_____

_____

Step 6:   **Use show commands to verify IPv6 settings.**

a. Check the status of the interfaces on R1 with the **show ipv6 interface brief** command.

What are the two IPv6 addresses for the G0/1 interface and what kind of IPv6 addresses are they?

_____

What are the two IPv6 addresses for the S0/0/1 interface and what kind of IPv6 addresses are they?

_____

b. To see more detailed information on the IPv6 interfaces, type a **show ipv6 interface** command on R1 and press Enter.

What are the multicast group addresses for the Gigabit Ethernet 0/1 interface?

_____

What are the multicast group addresses for the S0/0/1 interface?

_____

What is an FF02::1 multicast address used for?

_____

What is an FF02::2 multicast address used for?

_____

What kind of multicast addresses are FF02::1:FF00:1 and FF02::1:FF0D:1A60, and what are they used for?

_____

c. View the IPv6 routing table information for R1 using the **show ipv6 route** command. The IPv6 routing table should have two connected routes, one for each interface, and three local routes, one for each interface and one for multicast traffic to a Null0 interface.

In what way does the routing table output of R1 reveal why you were unable to ping PC-C from PC-A?

_____

_____

Part 2: Configure IPv6 Static and Default Routes

In Part 2, you will configure IPv6 static and default routes three different ways. You will confirm that the routes have been added to the routing tables, and you will verify successful connectivity between PC-A and PC-C.

You will configure three types of IPv6 static routes:

*   **Directly Connected IPv6 Static Route** – A directly connected static route is created when specifying the outgoing interface.

*   **Recursive IPv6 Static Route** – A recursive static route is created when specifying the next-hop IP address. This method requires the router to execute a recursive lookup in the routing table in order to identify the outgoing interface.

*   **Default IPv6 Static Route** – Similar to a quad zero IPv4 route, a default IPv6 static route is created by making the destination IPv6 prefix and prefix length all zeros, ::/0.

Step 1: Configure a directly connected IPv6 static route.

In a directly connected IPv6 static route, the route entry specifies the router outgoing interface. A directly connected static route is typically used with a point-to-point serial interface. To configure a directly attached IPv6 static route, use the following command format:

```
Router(config)# ipv6 route <ipv6-prefix/prefix-length> <outgoing-interface-type>
<outgoing-interface-number>
```

a. On router R1, configure an IPv6 static route to the 2001:DB8:ACAD:B::/64 network on R3, using the R1 outgoing S0/0/1 interface.

    ```
    R1(config)# ipv6 route 2001:DB8:ACAD:B::/64 serial 0/0/1
    R1(config)#
    ```

b. View the IPv6 routing table to verify the new static route entry.

    What is the code letter and routing table entry for the newly added route in the routing table?

    _____

    _____

c. Now that the static route has been configured on R1, is it now possible to ping the host PC-C from PC-A?

    _____

    These pings should fail. If the recursive static route is correctly configured, the ping arrives at PC-C. PC-C sends a ping reply back to PC-A. However, the ping reply is discarded at R3 because R3 does not have a return route to the 2001:DB8:ACAD:A::/64 network in the routing table. To successfully ping across the network, you must also create a static route on R3.

d. On router R3, configure an IPv6 static route to the 2001:DB8:ACAD:A::/64 network, using the R3 outgoing S0/0/0 interface.

    ```
    R3(config)# ipv6 route 2001:DB8:ACAD:A::/64 serial 0/0/0
    R3(config)#
    ```

e. Now that both routers have static routes, attempt an IPv6 **ping -6** from PC-A to the PC-C global unicast IPv6 address.

Was the ping successful? Why?

_____

Step 2:   Configure a recursive IPv6 static route.

In a recursive IPv6 static route, the route entry has the next-hop router IPv6 address. To configure a recursive IPv6 static route, use the following command format:

```
Router(config)# ipv6 route <ipv6-prefix/prefix-length> <next-hop-ipv6-address>
```

a.   On router R1, delete the directly attached static route and add a recursive static route.

```
R1(config)# no ipv6 route 2001:DB8:ACAD:B::/64 serial 0/0/1

R1(config)# ipv6 route 2001:DB8:ACAD:B::/64 FC00::2

R1(config)# exit
```

b.   On router R3, delete the directly attached static route and add a recursive static route.

```
R3(config)# no ipv6 route 2001:DB8:ACAD:A::/64 serial 0/0/0

R3(config)# ipv6 route 2001:DB8:ACAD:A::/64 FC00::1

R3(config)# exit
```

c.   View the IPv6 routing table on R1 to verify the new static route entry.

What is the code letter and routing table entry for the newly added route in the routing table?

_____

d. Verify connectivity by issuing a **ping -6** command from PC-A to PC-C.

Was the ping successful? _____

Note: It may be necessary to disable the PC firewall to ping between PCs.

Step 3: Configure a default IPv6 static route.

In a default static route, the destination IPv6 prefix and prefix length are all zeros.

```
Router(config)# ipv6 route ::/0 <outgoing-interface-type> <outgoing-interface-
number> {and/or} <next-hop-ipv6-address>
```

a. On router R1, delete the recursive static route and add a default static route.

```
R1(config)# no ipv6 route 2001:DB8:ACAD:B::/64 FC00::2

R1(config)# ipv6 route ::/0 serial 0/0/1

R1(config)#
```

b. Delete the recursive static route and add a default static route on R3.

c. View the IPv6 routing table on R1 to verify the new static route entry.

What is the code letter and routing table entry for the newly added default route in the routing table?

_____

d. Verify connectivity by issuing a **ping -6** command from PC-A to PC-C.

Was the ping successful? _____

Note: It may be necessary to disable the PC firewall to ping between PCs.

Reflection

1. This lab focuses on configuring IPv6 static and default routes. Can you think of a situation where you would need to configure both IPv6 and IPv4 static and default routes on a router?

_____

_____

_____

_____

2. In practice, configuring an IPv6 static and default route is very similar to configuring an IPv4 static and default route. Aside from the obvious differences between the IPv6 and IPv4 addressing, what are some other differences when configuring and verifying an IPv6 static route as compared to an IPv4 static route?

_____

_____

_____

Router Interface Summary Table

Router Interface Summary				
Router Model	Ethernet Interface #1	Ethernet Interface #2	Serial Interface #1	Serial Interface #2
1800	Fast Ethernet 0/0 (F0/0)	Fast Ethernet 0/1 (F0/1)	Serial 0/0/0 (S0/0/0)	Serial 0/0/1 (S0/0/1)
1900	Gigabit Ethernet 0/0 (G0/0)	Gigabit Ethernet 0/1 (G0/1)	Serial 0/0/0 (S0/0/0)	Serial 0/0/1 (S0/0/1)
2801	Fast Ethernet 0/0 (F0/0)	Fast Ethernet 0/1 (F0/1)	Serial 0/1/0 (S0/1/0)	Serial 0/1/1 (S0/1/1)
2811	Fast Ethernet 0/0 (F0/0)	Fast Ethernet 0/1 (F0/1)	Serial 0/0/0 (S0/0/0)	Serial 0/0/1 (S0/0/1)
2900	Gigabit Ethernet 0/0 (G0/0)	Gigabit Ethernet 0/1 (G0/1)	Serial 0/0/0 (S0/0/0)	Serial 0/0/1 (S0/0/1)
Note: To find out how the router is configured, look at the interfaces to identify the type of router and how many interfaces the router has. There is no way to effectively list all the combinations of configurations for each router class. This table includes identifiers for the possible combinations of Ethernet and Serial interfaces in the device. The table does not include any other type of interface, even though a specific router may contain one. An example of this might be an ISDN BRI interface. The string in parenthesis is the legal abbreviation that can be used in Cisco IOS commands to represent the interface.				

6.3.3.7 Lab – Designing and Implementing IPv4 Addressing with VLSM

Topology

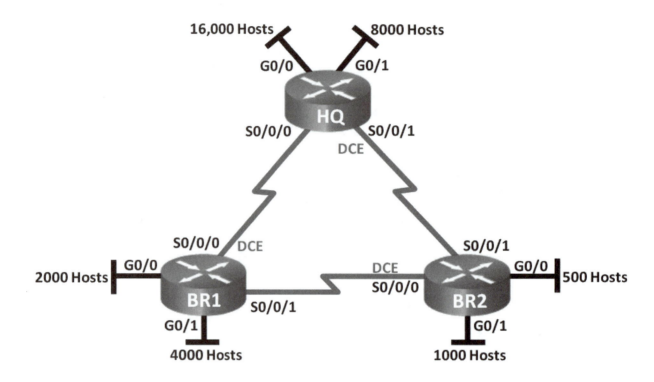

Objectives

Part 1: Examine the Network Requirements

Part 2: Design the VLSM Address Scheme

Part 3: Cable and Configure the IPv4 Network

Background / Scenario

The Variable Length Subnet Mask (VLSM) was designed to help conserve IP addresses. With VLSM, a network is subnetted and then subnetted again. This process can be repeated multiple times to create subnets of various sizes based on the number of hosts required in each subnet. Effective use of VLSM requires address planning.

In this lab, you are given the network address 172.16.128.0/17 to develop an address scheme for the network shown in the Topology diagram. VLSM will be used so that the addressing requirements can be met. After you have designed the VLSM address scheme, you will configure the interfaces on the routers with the appropriate IP address information.

Note: The routers used with CCNA hands-on labs are Cisco 1941 Integrated Services Routers (ISRs) with Cisco IOS Release 15.2(4)M3 (universalk9 image). Other routers and Cisco IOS versions can be used. Depending on the model and Cisco IOS version, the commands available and output produced might vary from what is shown in the labs. Refer to the Router Interface Summary Table at the end of this lab for the correct interface identifiers.

Note: Make sure that the routers have been erased and have no startup configurations. If you are unsure, contact your instructor.

Required Resources

- 3 Routers (Cisco 1941 with Cisco IOS Release 15.2(4)M3 universal image or comparable)
- 1 PC (with terminal emulation program, such as Tera Term, to configure routers)
- Console cable to configure the Cisco IOS devices via the console ports
- Ethernet (optional) and serial cables as shown in the topology
- Windows Calculator (optional)

Part 1: **Examine the Network Requirements**

In Part 1, you will examine the network requirements to develop a VLSM address scheme for the network shown in the Topology diagram using the network address of 172.16.128.0/17.

Note: You may use the Windows Calculator application and the www.ipcalc.org IP subnet calculator to help with your calculations.

Step 1: **Determine how many host addresses are available and how many subnets are needed.**

How many host addresses are available in a /17 network? _____

What is the total number of host addresses needed in the topology diagram? _____

How many subnets are needed in the network topology? _____

Step 2: **Determine the largest subnet needed.**

Subnet description (e.g. BR1 G0/1 LAN or BR1-HQ WAN link) _____

How many IP addresses are needed in the largest subnet? _____

What is the smallest subnet that supports that many addresses?

_____

How many host addresses does that subnet support? _____

Can the 172.16.128.0/17 network be subnetted to support this subnet? _____

What are the two network addresses that would result from this subnetting?

_____

_____

Use the first network address for this subnet.

Step 3: **Determine the second largest subnet needed.**

Subnet description _____

How many IP addresses are needed for the second largest subnet? _____

What is the smallest subnet that supports that many hosts?

_____

How many host addresses does that subnet support? _____

Can the remaining subnet be subnetted again and still support this subnet? _____

What are the two network addresses that would result from this subnetting?

_____

_____

Use the first network address for this subnet.

Step 4:   **Determine the next largest subnet needed.**

Subnet description _____

How many IP addresses are needed for the next largest subnet? _____

What is the smallest subnet that supports that many hosts?

_____

How many host addresses does that subnet support? _____

Can the remaining subnet be subnetted again and still support this subnet? _____

What are the two network addresses that would result from this subnetting?

_____

_____

Use the first network address for this subnet.

Step 5:   **Determine the next largest subnet needed.**

Subnet description _____

How many IP addresses are needed for the next largest subnet? _____

What is the smallest subnet that supports that many hosts?

_____

How many host addresses does that subnet support? _____

Can the remaining subnet be subnetted again and still support this subnet? _____

What are the two network addresses that would result from this subnetting?

_____

_____

Use the first network address for this subnet.

Step 6:   **Determine the next largest subnet needed.**

Subnet description _____

How many IP addresses are needed for the next largest subnet? _____

What is the smallest subnet that supports that many hosts?

_____

How many host addresses does that subnet support? _____

Can the remaining subnet be subnetted again and still support this subnet? _____

What are the two network addresses that would result from this subnetting?

_____

_____

Use the first network address for this subnet.

Step 7:   **Determine the next largest subnet needed.**

Subnet description _____

How many IP addresses are needed for the next largest subnet? _____

What is the smallest subnet that supports that many hosts?

_____

How many host addresses does that subnet support? _____

Can the remaining subnet be subnetted again and still support this subnet? _____

What are the two network addresses that would result from this subnetting?

_____

_____

Use the first network address for this subnet.

Step 8:   **Determine the subnets needed to support the serial links.**

How many host addresses are needed for each serial subnet link? _____

What is the smallest subnet that supports that many host addresses?

_____

a.   Subnet the remaining subnet and write the network addresses that result from this subnetting below.

_____

_____

b. Continue subnetting the first subnet of each new subnet until you have four /30 subnets. Write the first three network addresses of these /30 subnets below.

_____

_____

_____

c. Enter the subnet descriptions for these three subnets below.

_____

_____

_____

Part 2: **Design the VLSM Address Scheme**

Step 1: **Calculate the subnet information.**

Use the information that you obtained in Part 1 to fill in the table below.

Subnet Description	Number of Hosts Needed	Network Address /CIDR	First Host Address	Broadcast Address
HQ G0/0	16,000			
HQ G0/1	8,000			
BR1 G0/1	4,000			
BR1 G0/0	2,000			
BR2 G0/1	1,000			
BR2 G0/0	500			
HQ S0/0/0 – BR1 S0/0/0	2			
HQ S0/0/1 – BR2 S0/0/1	2			
BR1 S0/0/1 – BR2 S0/0/0	2			

Step 2: **Complete the device interface address table.**

Assign the first host address in the subnet to the Ethernet interfaces. HQ should be given the first host address on the serial links to BR1 and BR2. BR1 should be given the first host address for the serial link to BR2.

Device	Interface	IP Address	Subnet Mask	Device Interface
HQ	G0/0			16,000 Host LAN
	G0/1			8,000 Host LAN
	S0/0/0			BR1 S0/0/0
	S0/0/1			BR2 S0/0/1
BR1	G0/0			2,000 Host LAN
	G0/1			4,000 Host LAN
	S0/0/0			HQ S0/0/0
	S0/0/1			BR2 S0/0/0
BR2	G0/0			500 Host LAN
	G0/1			1,000 Host LAN
	S0/0/0			BR1 S0/0/1
	S0/0/1			HQ S0/0/1

Part 3: Cable and Configure the IPv4 Network

In Part 3, you will cable the network topology and configure the three routers using the VLSM address scheme that you developed in Part 2.

Step 1: Cable the network as shown in the topology.

Step 2: Configure basic settings on each router.

a. Assign the device name to the router.

b. Disable DNS lookup to prevent the router from attempting to translate incorrectly entered commands as though they were hostnames.

c. Assign **class** as the privileged EXEC encrypted password.

d. Assign **cisco** as the console password and enable login.

e. Assign **cisco** as the vty password and enable login.

f. Encrypt the clear text passwords.

g. Create a banner that warns anyone accessing the device that unauthorized access is prohibited.

Step 3: Configure the interfaces on each router.

a. Assign an IP address and subnet mask to each interface using the table that you completed in Part 2.

b. Configure an interface description for each interface.

c. Set the clocking rate on all DCE serial interfaces to 128000.

```
HQ(config-if)# clock rate 128000
```

d. Activate the interfaces.

Step 4:   **Save the configuration on all devices.**

Step 5:   **Test Connectivity.**

    a. From HQ, ping BR1's S0/0/0 interface address.

    b. From HQ, ping BR2's S0/0/1 interface address.

    c. From BR1, ping BR2's S0/0/0 interface address.

    d. Troubleshoot connectivity issues if pings were not successful.

Note: Pings to the GigabitEthernet interfaces on other routers are unsuccessful. The LANs defined for the GigabitEthernet interfaces are simulated. Because no devices are attached to these LANs, they are in a down/down state. A routing protocol must be in place for other devices to be aware of those subnets. The GigabitEthernet interfaces must also be in an up/up state before a routing protocol can add the subnets to the routing table. These interfaces remain in a down/down state until a device is connected to the other end of the Ethernet interface cable. The focus of this lab is on VLSM and configuring the interfaces.

Reflection

Can you think of a shortcut for calculating the network addresses of consecutive /30 subnets?

_____

_____

_____

_____

Router Interface Summary Table

Router Interface Summary				
Router Model	**Ethernet Interface #1**	**Ethernet Interface #2**	**Serial Interface #1**	**Serial Interface #2**
1800	Fast Ethernet 0/0 (F0/0)	Fast Ethernet 0/1 (F0/1)	Serial 0/0/0 (S0/0/0)	Serial 0/0/1 (S0/0/1)
1900	Gigabit Ethernet 0/0 (G0/0)	Gigabit Ethernet 0/1 (G0/1)	Serial 0/0/0 (S0/0/0)	Serial 0/0/1 (S0/0/1)
2801	Fast Ethernet 0/0 (F0/0)	Fast Ethernet 0/1 (F0/1)	Serial 0/1/0 (S0/1/0)	Serial 0/1/1 (S0/1/1)
2811	Fast Ethernet 0/0 (F0/0)	Fast Ethernet 0/1 (F0/1)	Serial 0/0/0 (S0/0/0)	Serial 0/0/1 (S0/0/1)
2900	Gigabit Ethernet 0/0 (G0/0)	Gigabit Ethernet 0/1 (G0/1)	Serial 0/0/0 (S0/0/0)	Serial 0/0/1 (S0/0/1)

Note: To find out how the router is configured, look at the interfaces to identify the type of router and how many interfaces the router has. There is no way to effectively list all the combinations of configurations for each router class. This table includes identifiers for the possible combinations of Ethernet and Serial interfaces in the device. The table does not include any other type of interface, even though a specific router may contain one. An example of this might be an ISDN BRI interface. The string in parenthesis is the legal abbreviation that can be used in Cisco IOS commands to represent the interface.

6.4.2.5 Lab – Calculating Summary Routes with IPv4 and IPv6

Topology

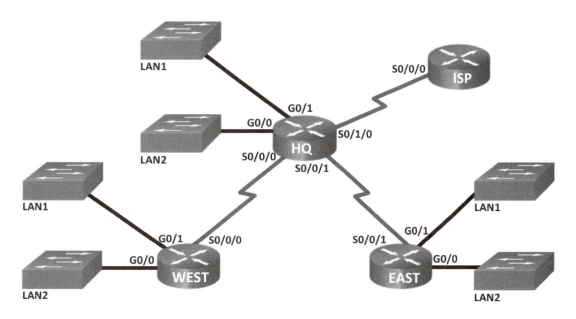

Addressing Table

Subnet	IPv4 Address	IPv6 Address
HQ LAN1	192.168.64.0/23	2001:DB8:ACAD:E::/64
HQ LAN2	192.168.66.0/23	2001:DB8:ACAD:F::/64
EAST LAN1	192.168.68.0/24	2001:DB8:ACAD:1::/64
EAST LAN2	192.168.69.0/24	2001:DB8:ACAD:2::/64
WEST LAN1	192.168.70.0/25	2001:DB8:ACAD:9::/64
WEST LAN2	192.168.70.128/25	2001:DB8:ACAD:A::/64
Link from HQ to EAST	192.168.71.4/30	2001:DB8:ACAD:1000::/64
Link from HQ to WEST	192.168.71.0/30	2001:DB8:ACAD:2000::/64
Link from HQ to ISP	209.165.201.0/30	2001:DB8:CC1E:1::/64

Objectives

Part 1: Calculate IPv4 Summary Routes

- Determine the summary route for the HQ LANs.
- Determine the summary route for the EAST LANs.
- Determine the summary route for the WEST LANs.
- Determine the summary route for the HQ, EAST, and WEST LANs.

Part 2: Calculate IPv6 Summary Routes

- Determine the summary route for the HQ LANs.

- Determine the summary route for the EAST LANs.

- Determine the summary route for the WEST LANs.

- Determine the summary route for the HQ, EAST, and WEST LANs.

Background / Scenario

Summary routes reduce the number of entries in routing tables and make the routing table lookup process more efficient. This process also reduces the memory requirements for the router. A single static route can be used to represent a few routes or thousands of routes.

In this lab, you will determine the summary routes for different subnets of a network. You will then determine the summary route for the entire network. Summary routes will be determined for both IPv4 and IPv6 addresses. Because IPv6 uses hexadecimal (hex) values, you will be required to convert hex to binary.

Required Resources

- 1 PC (Windows 7, Vista, or XP with Internet access)

- Optional: calculator for converting hex and decimal to binary

Part 1:   Calculate IPv4 Summary Routes

In Part 1, you will determine summarized routes that can be used to reduce the size of routing tables. Fill in the tables, after each set of steps, with the appropriate IPv4 addressing information.

Step 1:   **List the HQ LAN1 and HQ LAN2 IP subnet mask in decimal form.**

Step 2:   **List the HQ LAN1 and HQ LAN2 IP address in binary form.**

Step 3:   **Count the number of far left matching bits to determine the subnet mask for the summary route.**

a.   How many far left matching bits are present in the two networks? _____

b.   List the subnet mask for the summary route in decimal form.

Step 4:   **Copy the matching binary bits and then add all zeros to determine the summarized network address.**

a.   List the matching binary bits for HQ LAN1 and HQ LAN2 subnets.

b.   Add zeros to comprise the remainder of the network address in binary form.

c.   List the summarized network address in decimal form.

Subnet	IPv4 Address	Subnet Mask	Subnet IP Address in Binary Form
HQ LAN1	192.168.64.0		
HQ LAN2	192.168.66.0		
HQ LANs Summary Address			

Step 5:   List the EAST LAN1 and EAST LAN2 IP subnet mask in decimal form.

Step 6:   List the EAST LAN1 and EAST LAN2 IP address in binary form.

Step 7:   Count the number of far left matching bits to determine the subnet mask for the summary route.

a.   How many far left matching bits are present in the two networks? _____

b.   List the subnet mask for the summary route in decimal form.

Step 8:   Copy the matching binary bits and then add all zeros to determine the summarized network address.

a.   List the matching binary bits for EAST LAN1 and EAST LAN2 subnets.

b.   Add zeros to comprise the remainder of the network address in binary form.

c.   List the summarized network address in decimal form.

Subnet	IPv4 Address	Subnet Mask	Subnet Address in Binary Form
EAST LAN1	192.168.68.0		
EAST LAN2	192.168.69.0		
EAST LANs Summary Address			

Step 9:   List the WEST LAN1 and WEST LAN2 IP subnet mask in decimal form.

Step 10: List the WEST LAN1 and WEST LAN2 IP address in binary form.

Step 11: Count the number of far left matching bits to determine the subnet mask for the summary route.

a.   How many far left matching bits are present in the two networks? _____

b.   List the subnet mask for the summary route in decimal form.

Step 12: **Copy the matching binary bits and then add all zeros to determine the summarized network address.**

a. List the matching binary bits for WEST LAN1 and WEST LAN2 subnets.

b. Add zeros to comprise the remainder of the network address in binary form.

c. List the summarized network address in decimal form.

Subnet	IPv4 Address	Subnet Mask	Subnet IP Address in Binary Form
WEST LAN1	192.168.70.0		
WEST LAN2	192.168.70.128		
WEST LANs Summary Address			

Step 13: **List the HQ, EAST, and WEST summary route IP address and subnet mask in decimal form.**

Step 14: **List the HQ, EAST, and WEST summary route IP address in binary form.**

Step 15: **Count the number of far left matching bits to determine the subnet mask for the summary route.**

a. How many far left matching bits are present in the three networks? _____

b. List the subnet mask for the summary route in decimal form.

Step 16: **Copy the matching binary bits and then add all zeros to determine the summarized network address.**

a. List the matching binary bits for HQ, EAST, and WEST subnets.

b. Add zeros to comprise the remainder of the network address in binary form.

c. List the summarized network address in decimal form.

Subnet	IPv4 Address	Subnet Mask	Subnet IP Address in Binary Form
HQ			
EAST			
WEST			
Network Address Summary Route			

Part 2:   Calculate IPv6 Summary Routes

In Part 2, you will determine summarized routes that can be used to reduce the size of routing tables. Complete the tables after each set of steps, with the appropriate IPv6 addressing information.

Topology

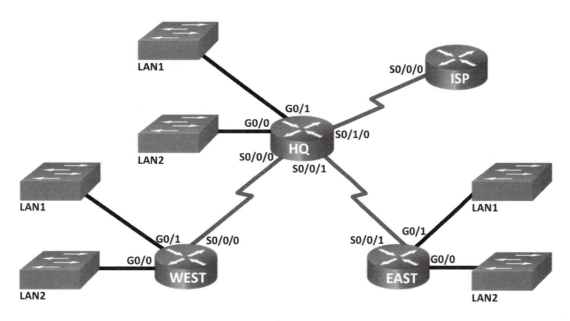

Addressing Table

Subnet	IPv6 Address
HQ LAN1	2001:DB8:ACAD:E::/64
HQ LAN2	2001:DB8:ACAD:F::/64
EAST LAN1	2001:DB8:ACAD:1::/64
EAST LAN2	2001:DB8:ACAD:2::/64
WEST LAN1	2001:DB8:ACAD:9::/64
WEST LAN2	2001:DB8:ACAD:A::/64
Link from HQ to EAST	2001:DB8:ACAD:1000::/64
Link from HQ to WEST	2001:DB8:ACAD:2000::/64
Link from HQ to ISP	2001:DB8:CC1E:1::/64

Step 1:   List the first 64 bits of the HQ LAN1 and HQ LAN2 IP subnet mask in hexadecimal form.

Step 2:   List the HQ LAN1 and HQ LAN2 subnet ID (bits 48-64) in binary form.

Step 3:   Count the number of far left matching bits to determine the subnet mask for the summary route.

a.   How many far left matching bits are present in the two subnet IDs? _____

b.   List the subnet mask for the first 64 bits of the summary route in decimal form.

Step 4:   **Copy the matching binary bits and then add all zeros to determine the summarized network address.**

   a.   List the matching subnet ID binary bits for HQ LAN1 and HQ LAN2 subnets.

   b.   Add zeros to comprise the remainder of the subnet ID address in binary form.

   c.   List the summarized network address in decimal form.

Subnet	IPv6 Address	Subnet Mask for First 64 bits	Subnet ID in Binary Form
HQ LAN1	2001:DB8:ACAD:E::/64		
HQ LAN2	2001:DB8:ACAD:F::/64		
HQ LANs Summary Address			

Step 5:   **List the first 64 bits of the EAST LAN1 and EAST LAN2 IP subnet mask in hexadecimal form.**

Step 6:   **List the EAST LAN1 and EAST LAN2 subnet ID (bits 48-64) in binary form.**

Step 7:   **Count the number of far left matching bits to determine the subnet mask for the summary route.**

   a.   How many far left matching bits are present in the two subnet IDs? _____

   b.   List the subnet mask for the first 64 bits of the summary route in decimal form.

Step 8:   **Copy the matching binary bits and then add all zeros to determine the summarized network address.**

   a.   List the matching binary bits for EAST LAN1 and EAST LAN2 subnets.

   b.   Add zeros to comprise the remainder of the subnet ID address in binary form.

   c.   List the summarized network address in decimal form.

Subnet	IPv6 Address	Subnet Mask for First 64 bits	Subnet ID in Binary Form
EAST LAN1	2001:DB8:ACAD:1::/64		
EAST LAN2	2001:DB8:ACAD:2::/64		
EAST LANs Summary Address			

Step 9:   **List the first 64 bits of the WEST LAN1 and WEST LAN2 IP subnet mask in decimal form.**

Step 10:   **List the WEST LAN1 and WEST LAN2 subnet ID (bits 48-64) in binary form.**

Step 11: **Count the number of far left matching bits to determine the subnet mask for the summary route.**

a. How many far left matching bits are present in the two subnet IDs? _____

b. List the subnet mask for the first 64 bits of the summary route in decimal form.

Step 12: **Copy the matching binary bits and then add all zeros to determine the summarized network address.**

a. List the matching binary bits for WEST LAN1 and WEST LAN2 subnets.

b. Add zeros to comprise the remainder of the subnet ID address in binary form.

c. List the summarized network address in decimal form.

Subnet	IPv6 Address	Subnet Mask for First 64 bits	Subnet ID in Binary Form
WEST LAN1	2001:DB8:ACAD:9::/64		
WEST LAN2	2001:DB8:ACAD:A::/64		
WEST LANs Summary Address			

Step 13: **List the HQ, EAST, and WEST summary route IP address and the first 64 bits of the subnet mask in decimal form.**

Step 14: **List the HQ, EAST, and WEST summary route subnet ID in binary form.**

Step 15: **Count the number of far left matching bits to determine the subnet mask for the summary route.**

a. How many far left matching bits are present in the three subnet IDs? _____

b. List the subnet mask for the first 64 bits of the summary route in decimal form.

Step 16: **Copy the matching binary bits and then add all zeros to determine the summarized network address.**

a. List the matching binary bits for HQ, EAST, and WEST subnets.

b. Add zeros to comprise the remainder of the subnet ID address in binary form.

c. List the summarized network address in decimal form.

Subnet	IPv6 Address	Subnet Mask for first 64 bits	Subnet ID in Binary Form
HQ			
EAST			
WEST			
Network Address Summary Route			

Reflection

1. How is determining the summary route for IPv4 different from IPv6?

_____

_____

_____

2. Why are summary routes beneficial to a network?

_____

_____

_____

6.5.2.5 Lab – Troubleshooting IPv4 and IPv6 Static Routes

Topology

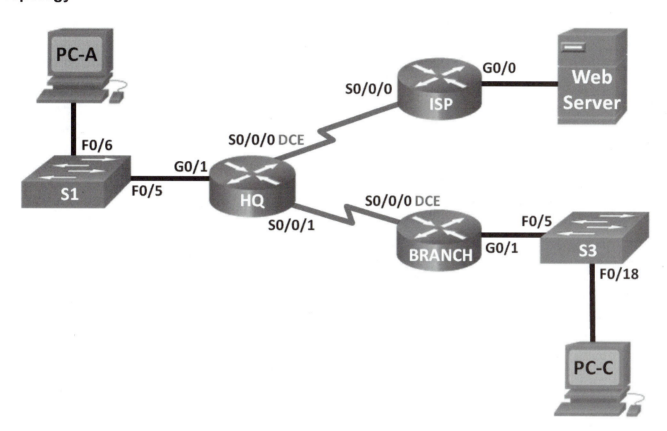

Addressing Table

Device	Interface	IP Address	Default Gateway
HQ	G0/1	192.168.0.1/25 2001:DB8:ACAD::1/64 FE80::1 link-local	N/A
	S0/0/0 (DCE)	10.1.1.2/30 2001:DB8:ACAD::20:2/64	N/A
	S0/0/1	192.168.0.253/30 2001:DB8:ACAD:2::1/30	N/A
ISP	G0/0	172.16.3.1/24 2001:DB8:ACAD:30::1/64 FE80::1 link-local	N/A
	S0/0/0	10.1.1.1/30 2001:DB8:ACAD:20::/64	N/A
BRANCH	G0/1	192.168.1.1/24 2001:DB8:ACAD:1::1/64 FE80::1 link-local	N/A
	S0/0/0 (DCE)	192.168.0.254/30 2001:DB8:ACAD:2::2/64	N/A
S1	VLAN 1	N/A	N/A
S3	VLAN 1	N/A	N/A
PC-A	NIC	192.168.0.3/25 2001:DB8:ACAD::3/64	192.168.0.1 FE80::1
Web Server	NIC	172.16.3.3/24 2001:DB8:ACAD:30::3/64	172.16.3.1 FE80::1
PC-C	NIC	192.168.1.3/24 2001:DB8:ACAD:1::3/64	192.168.1.1 FE80::1

Objectives

Part 1: Build the Network and Configure Basic Device Settings

Part 2: Troubleshoot Static Routes in an IPv4 Network

Part 3: Troubleshoot Static Routes in an IPv6 Network

Background / Scenario

As a network administrator, you must be able to configure routing of traffic using static routes. Understanding how to configure and troubleshoot static routing is a requirement. Static routes are commonly used for stub networks and default routes. Your company's ISP has hired you to troubleshoot connectivity issues on the network. You will have access to the HQ, BRANCH, and the ISP routers.

In this lab, you will begin by loading configuration scripts on each of the routers. These scripts contain errors that will prevent end-to-end communication across the network. You will need to troubleshoot each router to determine the configuration errors, and then use the appropriate commands to correct the configurations. When you have corrected all of the configuration errors, the hosts on the network should be able to communicate with each other.

Note: The routers used with CCNA hands-on labs are Cisco 1941 Integrated Services Routers (ISRs) with Cisco IOS Release 15.2(4)M3 (universalk9 image). The switches used are Cisco Catalyst 2960s with Cisco IOS Release 15.0(2) (lanbasek9 image). Other routers, switches, and Cisco IOS versions can be used. Depending on the model and Cisco IOS version, the commands available and output produced might vary from what is shown in the labs. Refer to the Router Interface Summary Table at the end of this lab for the correct interface identifiers.

Note: Make sure that the routers and switches have been erased and have no startup configurations. If you are unsure, contact your instructor.

Required Resources

- 3 Routers (Cisco 1941 with Cisco IOS Release 15.2(4)M3 universal image or comparable)
- 2 Switches (Cisco 2960 with Cisco IOS Release 15.0(2) lanbasek9 image or comparable)
- 3 PCs (Windows 7, Vista, or XP with terminal emulation program, such as Tera Term)
- Console cables to configure the Cisco IOS devices via the console ports
- Ethernet and serial cables as shown in the topology

Part 1: Build the Network and Configure Basic Device Settings

In Part 1, you will set up the network topology and configure the routers and switches with some basic settings, such as passwords and IP addresses. Preset configurations are also provided for you for the initial router configurations. You will also configure the IP settings for the PCs in the topology.

Step 1: Cable the network as shown in the topology.

Attach the devices as shown in the topology diagram and cable, as necessary.

Step 2: Initialize and reload the routers and switches.

Step 3: Configure basic settings for each router.

a. Disable DNS lookup.

b. Configure device name as shown in the topology.

c. Assign **class** as the privileged EXEC mode password.

d. Assign **cisco** as the console and vty passwords.

e. Configure **logging synchronous** to prevent console messages from interrupting command entry.

Step 4: Configure hosts and Web Server.

a. Configure IP addresses for IPv4 and IPv6.

b. Configure IPv4 default gateway.

Step 5: **Load router configurations.**

Router HQ

```
hostname HQ
ipv6 unicast-routing
interface GigabitEthernet0/1
 ipv6 address 2001:DB8:ACAD::1/64
 ip address 192.168.0.1 255.255.255.128
 ipv6 address FE80::1 link-local

interface Serial0/0/0
 ipv6 address 2001:DB8:ACAD:20::2/64
 ip address 10.1.1.2 255.255.255.252
 clock rate 800000
 no shutdown
interface Serial0/0/1
 ipv6 address 2001:DB8:ACAD:2::3/64

 ip address 192.168.0.253 255.255.255.252
 no shutdown
ip route 172.16.3.0 255.255.255.0 10.1.1.1
ip route 192.168.1.0 255.255.255.0 192.16.0.254

ipv6 route 2001:DB8:ACAD:1::/64 2001:DB8:ACAD:2::2
ipv6 route 2001:DB8:ACAD:30::/64 2001:DB8:ACAD::20:1
```

Router ISP

```
hostname ISP
ipv6 unicast-routing
interface GigabitEthernet0/0
 ipv6 address 2001:DB8:ACAD:30::1/64
 ip address 172.16.3.11 255.255.255.0

 ipv6 address FE80::1 link-local
 no shutdown
interface Serial0/0/0
 ipv6 address 2001:DB8::ACAD:20:1/64

 ip address 10.1.1.1 255.255.255.252
 no shutdown
```

```
ip route 192.168.1.0 255.255.255.0 10.1.1.2


ipv6 route 2001:DB8:ACAD::/62 2001:DB8:ACAD:20::2
```

Router BRANCH

```
hostname BRANCH

ipv6 unicast-routing

interface GigabitEthernet0/1

 ipv6 address 2001:DB8:ACAD:1::1/64

 ip address 192.168.1.1 255.255.255.0

 ipv6 address FE80::1 link-local

 no shutdown

interface Serial0/0/0

 ipv6 address 2001:DB8:ACAD:2::2/64

 clock rate 128000

 ip address 192.168.0.249 255.255.255.252


 clock rate 128000

 no shutdown

ip route 0.0.0.0 0.0.0.0 10.1.1.2



ipv6 route ::/0 2001:DB8:ACAD::1
```

Part 2:   Troubleshoot Static Routes in an IPv4 Network

IPv4 Addressing Table

Device	Interface	IP Address	Subnet Mask	Default Gateway
HQ	G0/1	192.168.0.1	255.255.255.0	N/A
	S0/0/0 (DCE)	10.1.1.2	255.255.255.252	N/A
	S0/0/1	192.168.0.253	255.255.255.252	N/A
ISP	G0/0	172.16.3.1	255.255.255.0	N/A
	S0/0/0	10.1.1.1	255.255.255.252	N/A
BRANCH	G0/1	192.168.1.1	255.255.255.0	N/A
	S0/0/0 (DCE)	192.168.0.254	255.255.255.252	N/A
S1	VLAN 1	192.168.0.11	255.255.255.128	192.168.0.1
S3	VLAN 1	192.168.1.11	255.255.255.0	192.168.1.1
PC-A	NIC	192.168.0.3	255.255.255.128	192.168.0.1
Web Server	NIC	172.16.3.3	255.255.255.0	172.16.3.1
PC-C	NIC	192.168.1.3	255.255.255.0	192.168.1.1

Step 1:   **Troubleshoot the HQ router.**

The HQ router is the link between the ISP router and the BRANCH router. The ISP router represents the outside network while the BRANCH router represents the corporate network. The HQ router is configured with static routes to ISP and BRANCH networks.

a.   Display the status of the interfaces on HQ. Enter **show ip interface brief**. Record and resolve any issues as necessary.

_____

_____

b.   Ping from HQ router to BRANCH router (192.168.0.254). Were the pings successful? _____

c.   Ping from HQ router to ISP router (10.1.1.1). Were the pings successful? _____

d.   Ping from PC-A to the default gateway. Were the pings successful? _____

e.   Ping from PC-A to PC-C. Were the pings successful? _____

f.   Ping from PC-A to Web Server. Were the pings successful? _____

g.   Display the routing table on HQ. What non-directly connected routes are shown in the routing table?

_____

_____

h.   Based on the results of the pings, routing table output, and static routes in the running configuration, what can you conclude about network connectivity?

_____

_____

_____

i. What commands (if any) need to be entered to resolve routing issues? Record the command(s).

_____

_____

_____

j. Repeat any of the steps from b to f to verify whether the problems have been resolved. Record your observations and possible next steps in troubleshooting connectivity.

_____

_____

_____

Step 2:   **Troubleshoot the ISP router.**

For the ISP router, there should be a route to HQ and BRANCH routers. One static route is configured on ISP router to reach the 192.168.1.0/24, 192.168.0.0/25, and 192.168.0.252/30 networks.

a. Display the status of interfaces on ISP. Enter **show ip interface brief**. Record and resolve any issues as necessary.

_____

_____

_____

b.   Ping from the ISP router to the HQ router (10.1.1.2). Were the pings successful? _____

c.   Ping from Web Server to the default gateway. Were the pings successful? _____

d.   Ping from Web Server to PC-A. Were the pings successful? _____

e.   Ping from Web Server to PC-C. Were the pings successful? _____

f.   Display the routing table on ISP. What non-directly connected routes are shown in the routing table?

_____

_____

g.   Based on the results of the pings, routing table output, and static routes in the running configuration, what can you conclude about network connectivity?

_____

_____

_____

h.   What commands (if any) need to be entered to resolve routing issues? Record the command(s).

(Hint: ISP only requires one summarized route to the company's networks 192.168.1.0/24, 192.168.0.0/25, and 192.168.0.252/32.)

_____

_____

_____

i. Repeat any of the steps from b to e to verify whether the problems have been resolved. Record your observations and possible next steps in troubleshooting connectivity.

_____

_____

_____

Step 3:   Troubleshoot the BRANCH router.

For the BRANCH router, a default route is set to reach the rest of the network and ISP.

a. Display the status of the interfaces on BRANCH. Enter **show ip interface brief**. Record and resolve any issues, as necessary.

_____

_____

b. Ping from the BRANCH router to the HQ router (192.168.0.253). Were the pings successful? _____

c.   Ping from PC-C to the default gateway. Were the pings successful? _____

d.   Ping from PC-C to PC-A. Were the pings successful? _____

e.   Ping from PC-C to Web Server. Were the pings successful? _____

f.   Display the routing table on BRANCH. What non-directly connected routes are shown in the routing table?

_____

_____

g.   Based on the results of the pings, routing table output, and static routes in the running configuration, what can you conclude about network connectivity?

_____

_____

_____

h.   What commands (if any) need to be entered to resolve routing issues? Record the command(s).

_____

_____

_____

i.   Repeat any of the steps from b to e to verify whether the problems have been resolved. Record your observations and possible next steps in troubleshooting connectivity.

_____

_____

_____

Part 3: **Troubleshoot Static Routes in an IPv6 Network**

Device	Interface	IPv6 Address	Prefix Length	Default Gateway
HQ	G0/1	2001:DB8:ACAD::1	64	N/A
	S0/0/0 (DCE)	2001:DB8:ACAD::20:2	64	N/A
	S0/0/1	2001:DB8:ACAD:2::1	64	N/A
ISP	G0/0	2001:DB8:ACAD:30::1	64	N/A
	S0/0/0	2001:DB8:ACAD:20::1	64	N/A
BRANCH	G0/1	2001:DB8:ACAD:1::1	64	N/A
	S0/0/0 (DCE)	2001:DB8:ACAD:2::2	64	N/A
PC-A	NIC	2001:DB8:ACAD::3	64	FE80::1
Web Server	NIC	2001:DB8:ACAD:30::3	64	FE80::1
PC-C	NIC	2001:DB8:ACAD:1::3	64	FE80::1

Step 1: **Troubleshoot the HQ router.**

The HQ router is the link between the ISP router and the BRANCH router. The ISP router represents the outside network while the BRANCH router represents the corporate network. The HQ router is configured with static routes to both the ISP and the BRANCH networks.

a. Display the status of the interfaces on HQ. Enter **show ipv6 interface brief**. Record and resolve any issues, as necessary.

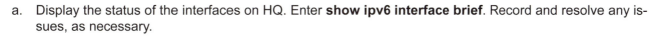

_____

_____

b. Ping from the HQ router to the BRANCH router (2001:DB8:ACAD:2::2). Were the pings successful?

_____

c. Ping from the HQ router to the ISP router (2001:DB8:ACAD:20::1). Were the pings successful? _____

d. Ping from PC-A to the default gateway. Were the pings successful? _____

e. Ping from PC-A to Web Server. Were the pings successful? _____

f. Ping from PC-A to PC-C. Were the pings successful? _____

g. Display the routing table by issuing a **show ipv6 route** command. What non-directly connected routes are shown in the routing table?

_____

_____

_____

h. Based on the results of the pings, routing table output, and static routes in the running configuration, what can you conclude about network connectivity?

_____

_____

_____

i. What commands (if any) need to be entered to resolve routing issues? Record the command(s).

_____

_____

_____

j. Repeat any of the steps from b to f to verify whether the problems have been resolved. Record your observations and possible next steps in troubleshooting connectivity.

_____

_____

_____

Step 2:   **Troubleshoot the ISP router.**

On the ISP router, one static route is configured to reach all the networks on HQ and BRANCH routers.

a. Display the status of the interfaces on ISP. Enter **show ipv6 interface brief**. Record and resolve any issues, as necessary.

_____

_____

b. Ping from the ISP router to the HQ router (2001:DB8:ACAD:20::2). Were the pings successful? _____

c. Ping from Web Server to the default gateway. Were the pings successful? _____

d. Ping from Web Server to PC-A. Were the pings successful? _____

e. Ping from Web Server to PC-C. Were the pings successful? _____

f. Display the routing table. What non-directly connected routes are shown in the routing table?

_____

_____

g. Based on the results of the pings, routing table output, and static routes in the running configuration, what can you conclude about network connectivity?

_____

_____

_____

h. What commands (if any) need to be entered to resolve routing issues? Record the command(s).

_____

_____

i. Repeat any of the steps from b to e to verify whether the problems have been resolved. Record your observations and possible next steps in troubleshooting connectivity.

_____

_____

_____

Step 3:   **Troubleshoot the BRANCH router.**

For the BRANCH routers, there is a default route to the HQ router. This default route allows the BRANCH network to the ISP router and Web Server.

a. Display the status of the interfaces on BRANCH. Enter **show ipv6 interface brief**. Record and resolve any issues, as necessary.

_____

_____

b. Ping from the BRANCH router to the HQ router (2001:DB8:ACAD:2::1). Were the pings successful?

_____

c. Ping from the BRANCH router to the ISP router (2001:DB8:ACAD:20::1). Were the pings successful?

_____

d. Ping from PC-C to the default gateway. Were the pings successful? _____

e. Ping from PC-C to PC-A. Were the pings successful? _____

f.   Ping from PC-C to Web Server. Were the pings successful? _____

g.   Display the routing table. What non-directly connected routes are shown in the routing table?

_____

_____

h.   Based on the results of the pings, routing table output, and static routes in the running configuration, what can
     you conclude about network connectivity?

_____

_____

_____

i.   What commands (if any) need to be entered to resolve routing issues? Record the command(s).

_____

_____

j.   Repeat any of the steps from b to f to verify whether the problems have been resolved. Record your ob-
     servations and possible next steps in troubleshooting connectivity.

_____

_____

_____

Router Interface Summary Table

Router Interface Summary				
Router Model	**Ethernet Interface #1**	**Ethernet Interface #2**	**Serial Interface #1**	**Serial Interface #2**
1800	Fast Ethernet 0/0 (F0/0)	Fast Ethernet 0/1 (F0/1)	Serial 0/0/0 (S0/0/0)	Serial 0/0/1 (S0/0/1)
1900	Gigabit Ethernet 0/0 (G0/0)	Gigabit Ethernet 0/1 (G0/1)	Serial 0/0/0 (S0/0/0)	Serial 0/0/1 (S0/0/1)
2801	Fast Ethernet 0/0 (F0/0)	Fast Ethernet 0/1 (F0/1)	Serial 0/1/0 (S0/1/0)	Serial 0/1/1 (S0/1/1)
2811	Fast Ethernet 0/0 (F0/0)	Fast Ethernet 0/1 (F0/1)	Serial 0/0/0 (S0/0/0)	Serial 0/0/1 (S0/0/1)
2900	Gigabit Ethernet 0/0 (G0/0)	Gigabit Ethernet 0/1 (G0/1)	Serial 0/0/0 (S0/0/0)	Serial 0/0/1 (S0/0/1)

Note: To find out how the router is configured, look at the interfaces to identify the type of router and how many interfaces the router has. There is no way to effectively list all the combinations of configurations for each router class. This table includes identifiers for the possible combinations of Ethernet and Serial interfaces in the device. The table does not include any other type of interface, even though a specific router may contain one. An example of this might be an ISDN BRI interface. The string in parenthesis is the legal abbreviation that can be used in Cisco IOS commands to represent the interface.

6.6.1.1 Class Activity – Make It Static!

Objectives

Configure a static route.

As the use of IPv6 addressing becomes more prevalent, it is important for network administrators to be able to direct network traffic between routers.

To prove that you are able to direct IPv6 traffic correctly and review the IPv6 default static route curriculum concepts, use the topology as shown in the .pdf file provided, specifically for this activity. Work with a partner to write an IPv6 statement for each of the three scenarios. Try to write the route statements without the assistance of completed labs, Packet Tracer files, etc.

- **Scenario 1**

    IPv6 default static route from R2 directing all data through your S0/0/0 interface to the next hop address on R1.

- **Scenario 2**

    IPv6 default static route from R3 directing all data through your S0/0/1 interface to the next hop address on R2.

- **Scenario 3**

    IPv6 default static route from R2 directing all data through your S0/0/1 interface to the next hop address on R3.

When complete, get together with another group and compare your written answers. Discuss any differences found in your comparisons.

Resources

Topology Diagram

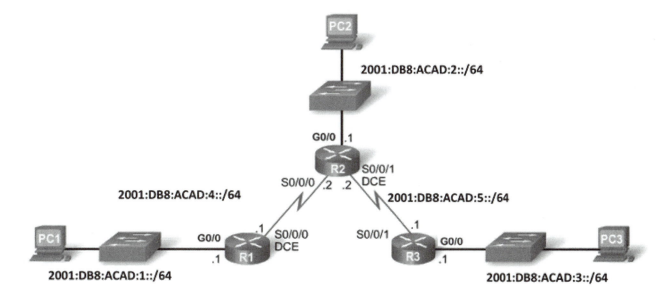

- **Scenario 1**

  IPv6 default static route from R2 directing all data to the next hop address on R1.

Configuration Command	IPv6 Network to Route	Next Hop IPv6 Address
R2(config)# **ipv6 route**		

- **Scenario 2**

  IPv6 default static route from R3 directing all data to the next hop address on R2.

Configuration Command	IPv6 Network to Route	Next Hop IPv6 Address
R3(config)# **ipv6 route**		

- **Scenario 3**

  IPv6 default static route from R2 directing all data to the next hop address on R3.

Configuration Command	IPv6 Network to Route	Next Hop IPv6 Address
R2(config)# **ipv6 route**		

Chapter 7 — Routing Dynamically

7.0.1.2 Class Activity – How Much Does This Cost?

Objectives

Explain the operation of dynamic routing protocols.

Scenario

This modeling activity illustrates the network concept of routing cost.

You will be a member of a team of five students who travel routes to complete the activity scenarios. One digital camera or bring your own device (BYOD) with camera, a stopwatch, and the student file for this activity will be required per group. One person will function as the photographer and event recorder, as selected by each group. The remaining four team members will actively participate in the scenarios below.

A school or university classroom, hallway, outdoor track area, school parking lot, or any other location can serve as the venue for these activities.

Activity 1

The tallest person in the group establishes a start and finish line by marking 15 steps from start to finish, indicating the distance of the team route. Each student will take 15 steps from the start line toward the finish line and then stop on the 15th step—no further steps are allowed.

Note: Not all of the students may reach the same distance from the start line due to their height and stride differences. The photographer will take a group picture of the entire team's final location after taking the 15 steps required.

Activity 2

A new start and finish line will be established; however, this time, a longer distance for the route will be established than the distance specified in Activity 1. No maximum steps are to be used as a basis for creating this particular route. One at a time, students will "walk the new route from beginning to end twice".

Each team member will count the steps taken to complete the route. The recorder will time each student and at the end of each team member's route, record the time that it took to complete the full route and how many steps were taken, as recounted by each team member and recorded on the team's student file.

Once both activities have been completed, teams will use the digital picture taken for Activity 1 and their recorded data from Activity 2 file to answer the reflection questions.

Group answers can be discussed as a class, time permitting.

Required Resources

- Digital or BYOD camera to record Activity 1's team results. Activity 2's data is based solely upon number of steps taken and the time it took to complete the route and no camera is necessary for Activity 2.
- Stopwatch
- Student file accompanying this modeling activity so that Activity 2 results can be recorded as each student finishes the route.

Scenario – Part 2 Recording Matrix

Student Team Member Name	Time Used to Finish the Route	Number of Steps Taken to Finish the Route

Reflection Questions

1. The photographer took a picture of the team's progress after taking 15 steps for Activity 1. Most likely, some team members did not reach the finish line on their 15th step due to height and stride differences. What do you think would happen if network data did not reach the finish line, or destination, in the allowed number of hops or steps?

_____

2. What could be done to help team members reach the finish line if they did not reach it in Activity 1?

_____

3. Which person would best be selected to deliver data using the network route completed in Activity 2? Justify your answer.

_____

4. Using the data recorded in Activity 2 and a limit of 255 steps, or hops, did all members of the team take more than 255 steps to finish their route? What would happen if they had to stop on the 254th step, or hop?

_____

5. Use the data that was recorded in Activity 2. Would you say the parameters for the route were enough to finish it successfully if all team members reached the finish line with 255 or less steps, or hops? Justify your answer.

_____

6. In network routing, different parameters are set for routing protocols. Use the data recorded for Activity 2. Would you select time, or number of steps, or hops, or a combination of both as your preferred routing type? List at least three reasons for your answers.

_____

7.3.2.4 Lab – Configuring Basic RIPv2 and RIPng

Topology

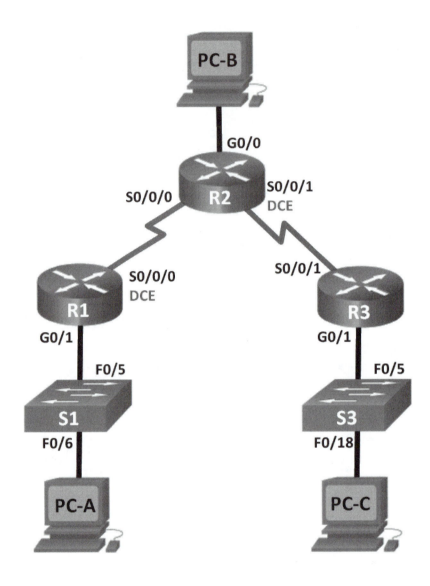

Addressing Table

Device	Interface	IP Address	Subnet Mask	Default Gateway
R1	G0/1	172.30.10.1	255.255.255.0	N/A
	S0/0/0 (DCE)	10.1.1.1	255.255.255.252	N/A
R2	G0/0	209.165.201.1	255.255.255.0	N/A
	S0/0/0	10.1.1.2	255.255.255.252	N/A
	S0/0/1 (DCE)	10.2.2.2	255.255.255.252	N/A
R3	G0/1	172.30.30.1	255.255.255.0	N/A
	S0/0/1	10.2.2.1	255.255.255.252	N/A
S1	N/A	VLAN 1	N/A	N/A
S3	N/A	VLAN 1	N/A	N/A
PC-A	NIC	172.30.10.3	255.255.255.0	172.30.10.1
PC-B	NIC	209.165.201.2	255.255.255.0	209.165.201.1
PC-C	NIC	172.30.30.3	255.255.255.0	172.30.30.1

Objectives

Part 1: Build the Network and Configure Basic Device Settings

Part 2: Configure and Verify RIPv2 Routing

- Configure and verify RIPv2 is running on routers.
- Configure a passive interface.
- Examine routing tables.
- Disable automatic summarization.
- Configure a default route.
- Verify end-to-end connectivity.

Part 3: Configure IPv6 on Devices

Part 4: Configure and Verify RIPng Routing

- Configure and verify RIPng is running on routers.
- Examine routing tables.
- Configure a default route.
- Verify end-to-end connectivity.

Background / Scenario

RIP version 2 (RIPv2) is used for routing of IPv4 addresses in small networks. RIPv2 is a classless, distance-vector routing protocol, as defined by RFC 1723. Because RIPv2 is a classless routing protocol, subnet masks are included in the routing updates. By default, RIPv2 automatically summarizes networks at major network boundaries. When automatic summarization has been disabled, RIPv2 no longer summarizes networks to their classful address at boundary routers.

RIPng (RIP Next Generation) is a distance-vector routing protocol for routing IPv6 addresses, as defined by RFC 2080. RIPng is based on RIPv2 and has the same administrative distance and 15-hop limitation.

In this lab, you will configure the network topology with RIPv2 routing, disable automatic summarization, propagate a default route, and use CLI commands to display and verify RIP routing information. You will then configure the network topology with IPv6 addresses, configure RIPng, propagate a default route, and use CLI commands to display and verify RIPng routing information.

Note: The routers used with CCNA hands-on labs are Cisco 1941 Integrated Services Routers (ISRs) with Cisco IOS Release 15.2(4)M3 (universalk9 image). The switches used are Cisco Catalyst 2960s with Cisco IOS Release 15.0(2) (lanbasek9 image). Other routers, switches, and Cisco IOS versions can be used. Depending on the model and Cisco IOS version, the commands available and output produced might vary from what is shown in the labs. Refer to the Router Interface Summary Table at the end of the lab for the correct interface identifiers.

Note: Make sure that the routers and switches have been erased and have no startup configurations. If you are unsure, contact your instructor.

Required Resources

- 3 Routers (Cisco 1941 with Cisco IOS Release 15.2(4)M3 universal image or comparable)
- 2 Switches (Cisco 2960 with Cisco IOS Release 15.0(2) lanbasek9 image or comparable)
- 3 PCs (Windows 7, Vista, or XP with terminal emulation program, such as Tera Term)
- Console cables to configure the Cisco IOS devices via the console ports
- Ethernet and Serial cables as shown in the topology

Part 1: Build the Network and Configure Basic Device Settings

In Part 1, you will set up the network topology and configure basic settings.

Step 1: **Cable the network as shown in the topology.**

Step 2: **Initialize and reload the router and switch.**

Step 3: **Configure basic settings for each router and switch.**

a. Disable DNS lookup.

b. Configure device names as shown in the topology.

c. Configure password encryption.

d. Assign **class** as the privileged EXEC password.

e. Assign **cisco** as the console and vty passwords.

f. Configure a MOTD banner to warn users that unauthorized access is prohibited.

g. Configure **logging synchronous** for the console line.

h. Configure the IP address listed in the Addressing Table for all interfaces.

i. Configure a description to each interface with an IP address.

j. Configure the clock rate if applicable to the DCE serial interface.

k. Copy the running-configuration to the startup-configuration.

Step 4: Configure PC hosts.

Refer to the Addressing Table for PC host address information.

Step 5: Test connectivity.

At this point, the PCs are unable to ping each other.

a. Each workstation should be able to ping the attached router. Verify and troubleshoot if necessary.

b. The routers should be able to ping one another. Verify and troubleshoot if necessary.

Part 2: Configure and Verify RIPv2 Routing

In Part 2, you will configure RIPv2 routing on all routers in the network and then verify that routing tables are updated correctly. After RIPv2 has been verified, you will disable automatic summarization, configure a default route, and verify end-to-end connectivity.

Step 1: Configure RIPv2 routing.

a. On R1, configure RIPv2 as the routing protocol and advertise the appropriate networks.

```
R1# config t
R1(config)# router rip
R1(config-router)# version 2
R1(config-router)# passive-interface g0/1
R1(config-router)# network 172.30.0.0
R1(config-router)# network 10.0.0.0
```

The **passive-interface** command stops routing updates out the specified interface. This process prevents unnecessary routing traffic on the LAN. However, the network that the specified interface belongs to is still advertised in routing updates that are sent out across other interfaces.

b. Configure RIPv2 on R3 and use the **network** statement to add appropriate networks and prevent routing updates on the LAN interface.

c. Configure RIPv2 on R2. Do not advertise the 209.165.201.0 network.

Note: It is not necessary to make the G0/0 interface passive on R2 because the network associated with this interface is not being advertised.

Step 2: Examine current state of network.

a. The status of the two serial links can quickly be verified using the **show ip interface brief** command on R2.

```
R2# show ip interface brief
Interface                  IP-Address      OK? Method Status                Protocol
Embedded-Service-Engine0/0 unassigned      YES unset  administratively down down
```

```
GigabitEthernet0/0          209.165.201.1    YES manual up                          up
GigabitEthernet0/1          unassigned       YES unset  administratively down down
Serial0/0/0                 10.1.1.2         YES manual up                          up
Serial0/0/1                 10.2.2.2         YES manual up                          up
```

b. Check connectivity between PCs.

From PC-A, is it possible to ping PC-B? _____ Why?

From PC-A, is it possible to ping PC-C? _____ Why?

From PC-C, is it possible to ping PC-B? _____ Why?

From PC-C, is it possible to ping PC-A? _____ Why?

c. Verify that RIPv2 is running on the routers.

You can use the **debug ip rip**, **show ip protocols**, and **show run** commands to confirm that RIPv2 is running. The **show ip protocols** command output for R1 is shown below.

```
R1# show ip protocols
Routing Protocol is "rip"
Outgoing update filter list for all interfaces is not set
Incoming update filter list for all interfaces is not set
Sending updates every 30 seconds, next due in 7 seconds
Invalid after 180 seconds, hold down 180, flushed after 240
Redistributing: rip
Default version control: send version 2, receive 2
  Interface            Send  Recv  Triggered RIP  Key-chain
  Serial0/0/0            2     2
Automatic network summarization is in effect
Maximum path: 4
Routing for Networks:
  10.0.0.0
  172.30.0.0
Passive Interface(s):
    GigabitEthernet0/1
Routing Information Sources:
  Gateway          Distance      Last Update
  10.1.1.2              120
Distance: (default is 120)
```

When issuing the **debug ip rip** command on R2, what information is provided that confirms RIPv2 is running?

_____

_____

When you are finished observing the debugging outputs, issue the **undebug all** command at the privileged EXEC prompt.

When issuing the **show run** command on R3, what information is provided that confirms RIPv2 is running?

_____

_____

d. Examine the automatic summarization of routes.

The LANs connected to R1 and R3 are composed of discontiguous networks. R2 displays two equal-cost paths to the 172.30.0.0/16 network in the routing table. R2 displays only the major classful network address of 172.30.0.0 and does not display any of the subnets for this network.

R2# **show ip route**

```
<Output omitted>

      10.0.0.0/8 is variably subnetted, 4 subnets, 2 masks
C        10.1.1.0/30 is directly connected, Serial0/0/0
L        10.1.1.2/32 is directly connected, Serial0/0/0
C        10.2.2.0/30 is directly connected, Serial0/0/1
L        10.2.2.2/32 is directly connected, Serial0/0/1
R     172.30.0.0/16 [120/1] via 10.2.2.1, 00:00:23, Serial0/0/1
                    [120/1] via 10.1.1.1, 00:00:09, Serial0/0/0
      209.165.201.0/24 is variably subnetted, 2 subnets, 2 masks
C        209.165.201.0/24 is directly connected, GigabitEthernet0/0
L        209.165.201.1/32 is directly connected, GigabitEthernet0/0
```

R1 displays only its own subnets for the 172.30.0.0 network. R1 does not have any routes for the 172.30.0.0 subnets on R3.

R1# **show ip route**

```
<Output omitted>

      10.0.0.0/8 is variably subnetted, 3 subnets, 2 masks
C        10.1.1.0/30 is directly connected, Serial0/0/0
L        10.1.1.1/32 is directly connected, Serial0/0/0
R        10.2.2.0/30 [120/1] via 10.1.1.2, 00:00:21, Serial0/0/0
      172.30.0.0/16 is variably subnetted, 2 subnets, 2 masks
C        172.30.10.0/24 is directly connected, GigabitEthernet0/1
L        172.30.10.1/32 is directly connected, GigabitEthernet0/1
```

R3 only displays its own subnets for the 172.30.0.0 network. R3 does not have any routes for the 172.30.0.0 subnets on R1.

```
R3# show ip route

<Output omitted>

        10.0.0.0/8 is variably subnetted, 3 subnets, 2 masks
C          10.2.2.0/30 is directly connected, Serial0/0/1
L          10.2.2.1/32 is directly connected, Serial0/0/1
R          10.1.1.0/30 [120/1] via 10.2.2.2, 00:00:23, Serial0/0/1
        172.30.0.0/16 is variably subnetted, 2 subnets, 2 masks
C          172.30.30.0/24 is directly connected, GigabitEthernet0/1
L          172.30.30.1/32 is directly connected, GigabitEthernet0/1
```

Use the **debug ip rip** command on R2 to determine the routes received in the RIP updates from R3 and list them here.

_____

R3 is not sending any of the 172.30.0.0 subnets, only the summarized route of 172.30.0.0/16, including the subnet mask. Therefore, the routing tables on R1 and R2 do not display the 172.30.0.0 subnets on R3.

Step 3:   Disable automatic summarization.

a.   The **no auto-summary** command is used to turn off automatic summarization in RIPv2. Disable auto summarization on all routers. The routers will no longer summarize routes at major classful network boundaries. R1 is shown here as an example.

```
R1(config)# router rip

R1(config-router)# no auto-summary
```

b.   Issue the **clear ip route** * command to clear the routing table.

```
R1(config-router)# end

R1# clear ip route *
```

c.   Examine the routing tables. Remember will it take some time to converge the routing tables after clearing them.

The LAN subnets connected to R1 and R3 should now be included in all three routing tables.

```
R2# show ip route

<Output omitted>

Gateway of last resort is not set


        10.0.0.0/8 is variably subnetted, 4 subnets, 2 masks
C          10.1.1.0/30 is directly connected, Serial0/0/0
L          10.1.1.2/32 is directly connected, Serial0/0/0
C          10.2.2.0/30 is directly connected, Serial0/0/1
L          10.2.2.2/32 is directly connected, Serial0/0/1
        172.30.0.0/16 is variably subnetted, 3 subnets, 2 masks
R          172.30.0.0/16 [120/1] via 10.2.2.1, 00:01:01, Serial0/0/1
                         [120/1] via 10.1.1.1, 00:01:15, Serial0/0/0
```

```
R         172.30.10.0/24 [120/1] via 10.1.1.1, 00:00:21, Serial0/0/0
R         172.30.30.0/24 [120/1] via 10.2.2.1, 00:00:04, Serial0/0/1
       209.165.201.0/24 is variably subnetted, 2 subnets, 2 masks
C         209.165.201.0/24 is directly connected, GigabitEthernet0/0
L         209.165.201.1/32 is directly connected, GigabitEthernet0/0
```

```
R1# show ip route
<Output omitted>
Gateway of last resort is not set


       10.0.0.0/8 is variably subnetted, 3 subnets, 2 masks
C         10.1.1.0/30 is directly connected, Serial0/0/0
L         10.1.1.1/32 is directly connected, Serial0/0/0
R         10.2.2.0/30 [120/1] via 10.1.1.2, 00:00:12, Serial0/0/0
       172.30.0.0/16 is variably subnetted, 3 subnets, 2 masks
C         172.30.10.0/24 is directly connected, GigabitEthernet0/1
L         172.30.10.1/32 is directly connected, GigabitEthernet0/1
R         172.30.30.0/24 [120/2] via 10.1.1.2, 00:00:12, Serial0/0/0
```

```
R3# show ip route
<Output omitted>
       10.0.0.0/8 is variably subnetted, 3 subnets, 2 masks
C         10.2.2.0/30 is directly connected, Serial0/0/1
L         10.2.2.1/32 is directly connected, Serial0/0/1
R         10.1.1.0/30 [120/1] via 10.2.2.2, 00:00:23, Serial0/0/1
       172.30.0.0/16 is variably subnetted, 2 subnets, 2 masks
C         172.30.30.0/24 is directly connected, GigabitEthernet0/1
L         172.30.30.1/32 is directly connected, GigabitEthernet0/1
R         172.30.10.0 [120/2] via 10.2.2.2, 00:00:16, Serial0/0/1
```

d. Use the **debug ip rip** command on R2 to exam the RIP updates.

```
R2# debug ip rip
```

After 60 seconds, issue the **no debug ip rip** command.

What routes are in the RIP updates that are received from R3?

_____

Are the subnet masks now included in the routing updates? _____

Step 4: **Configure and redistribute a default route for Internet access.**

a. From R2, create a static route to network 0.0.0.0 0.0.0.0, using the **ip route** command. This forwards any unknown destination address traffic to the R2 G0/0 toward PC-B, simulating the Internet by setting a Gateway of Last Resort on the R2 router.

```
R2(config)# ip route 0.0.0.0 0.0.0.0 209.165.201.2
```

b. R2 will advertise a route to the other routers if the **default-information originate** command is added to its RIP configuration.

```
R2(config)# router rip

R2(config-router)# default-information originate
```

Step 5:   Verify the routing configuration.

a. View the routing table on R1.

```
R1# show ip route

<Output omitted>

Gateway of last resort is 10.1.1.2 to network 0.0.0.0


R*      0.0.0.0/0 [120/1] via 10.1.1.2, 00:00:13, Serial0/0/0

        10.0.0.0/8 is variably subnetted, 3 subnets, 2 masks

C          10.1.1.0/30 is directly connected, Serial0/0/0

L          10.1.1.1/32 is directly connected, Serial0/0/0

R          10.2.2.0/30 [120/1] via 10.1.1.2, 00:00:13, Serial0/0/0

        172.30.0.0/16 is variably subnetted, 3 subnets, 2 masks

C          172.30.10.0/24 is directly connected, GigabitEthernet0/1

L          172.30.10.1/32 is directly connected, GigabitEthernet0/1

R          172.30.30.0/24 [120/2] via 10.1.1.2, 00:00:13, Serial0/0/0
```

How can you tell from the routing table that the subnetted network shared by R1 and R3 has a pathway for Internet traffic?

_____

_____

b. View the routing table on R2.

How is the pathway for Internet traffic provided in its routing table?

_____

_____

Step 6:   Verify connectivity.

a. Simulate sending traffic to the Internet by pinging from PC-A and PC-C to 209.165.201.2.

Were the pings successful? _____

b. Verify that hosts within the subnetted network can reach each other by pinging between PC-A and PC-C.

Were the pings successful? _____

Note: It may be necessary to disable the PCs firewall.

Part 3:   **Configure IPv6 on Devices**

In Part 3, you will configure all interfaces with IPv6 addresses and verify connectivity.

Addressing Table

Device	Interface	IPv6 Address / Prefix Length	Default Gateway
R1	G0/1	2001:DB8:ACAD:A::1/64	
		FE80::1 link-local	N/A
	S0/0/0	2001:DB8:ACAD:12::1/64	
		FE80::1 link-local	N/A
R2	G0/0	2001:DB8:ACAD:B::2/64	
		FE80::2 link-local	N/A
	S0/0/0	2001:DB8:ACAD:12::2/64	
		FE80::2 link-local	N/A
	S0/0/1	2001:DB8:ACAD:23::2/64	
		FE80::2 link-local	N/A
R3	G0/1	2001:DB8:ACAD:C::3/64	
		FE80::3 link-local	N/A
	S0/0/1	2001:DB8:ACAD:23::3/64	
		FE80::3 link-local	N/A
PC-A	NIC	2001:DB8:ACAD:A::A/64	FE80::1
PC-B	NIC	2001:DB8:ACAD:B::B/64	FE80::2
PC-C	NIC	2001:DB8:ACAD:C::C/64	FE80::3

Step 1:   **Configure PC hosts.**

Refer to the Addressing Table for PC host address information.

Step 2:   **Configure IPv6 on routers.**

Note: Assigning an IPv6 address in addition to an IPv4 address on an interface is known as dual stacking. This is because both IPv4 and IPv6 protocol stacks are active.

a.   For each router interface, assign the global and link local address from the Addressing Table.

b.   Enable IPv6 routing on each router.

c.   Enter the appropriate command to verify IPv6 addresses and link status. Write the command in the space below.

d.   Each workstation should be able to ping the attached router. Verify and troubleshoot if necessary.

e.   The routers should be able to ping one another. Verify and troubleshoot if necessary.

Part 4:   Configure and Verify RIPng Routing

In Part 4, you will configure RIPng routing on all routers, verify that routing tables are updated correctly, configure and distribute a default route, and verify end-to-end connectivity.

Step 1:   Configure RIPng routing.

With IPv6, it is common to have multiple IPv6 addresses configured on an interface. The network statement has been eliminated in RIPng. RIPng routing is enabled at the interface level instead, and is identified by a locally significant process name as multiple processes can be created with RIPng.

a.   Issue the **ipv6 rip Test1 enable** command for each interface on R1 that is to participate in RIPng routing, where **Test1** is the locally significant process name.

```
R1(config)# interface g0/1
R1(config)# ipv6 rip Test1 enable
R1(config)# interface s0/0/0
R1(config)# ipv6 rip Test1 enable
```

b.   Configure RIPng for the serial interfaces on R2 with **Test2** as the process name. Do not configure for the G0/0 interface.

c.   Configure RIPng for each interface on R3 with **Test3** as the process name.

d.   Verify that RIPng is running on the routers.

The **show ipv6 protocols, show run, show ipv6 rip database**, and **show ipv6 rip** *process name* commands can all be used to confirm that RIPng is running. On R1, issue the **show ipv6 protocols** command.

```
R1# show ipv6 protocols
IPv6 Routing Protocol is "connected"
IPv6 Routing Protocol is "ND"
IPv6 Routing Protocol is "rip Test1"
  Interfaces:
    Serial0/0/0
    GigabitEthernet0/1
  Redistribution:
    None
```

How is the RIPng listed in the output?

_____

_____

e. Issue the **show ipv6 rip Test1** command.

    R1# **show ipv6 rip Test1**

    RIP process "Test1", port 521, multicast-group FF02::9, pid 314

          Administrative distance is 120. Maximum paths is 16

          Updates every 30 seconds, expire after 180

          Holddown lasts 0 seconds, garbage collect after 120

          Split horizon is on; poison reverse is off

          Default routes are not generated

          Periodic updates 1, trigger updates 0

          Full Advertisement 0, Delayed Events 0

       Interfaces:

         GigabitEthernet0/1

         Serial0/0/0

       Redistribution:

         None

    How are RIPv2 and RIPng similar?

    _____

    _____

f. Inspect the IPv6 routing table on each router. Write the appropriate command used to view the routing
    table in the space below.

    _____

    On R1, how many routes have been learned by RIPng? _____

    On R2, how many routes have been learned by RIPng? _____

    On R3, how many routes have been learned by RIPng? _____

g. Check connectivity between PCs.

From PC-A, is it possible to ping PC-B? _____

From PC-A, is it possible to ping PC-C? _____

From PC-C, is it possible to ping PC-B? _____

From PC-C, is it possible to ping PC-A? _____

Why are some pings successful and others not?

_____

Step 2: Configure and redistribute a default route.

a. From R2, create a static default route to network ::0/64 using the **ipv6 route** command, and the IP address of exit interface G0/0. This forwards any unknown destination address traffic to the R2 G0/0 interface toward PC-Bs, simulating the Internet. Write the command used in the space below.

_____

b. Static routes can be included in RIPng updates by using the **ipv6 rip** *process name* **default-information originate** command in interface configuration mode. Configure the serial links on R2 to send the default route in RIPng updates.

```
R2(config)# int s0/0/0

R2(config-rtr)# ipv6 rip Test2 default-information originate

R2(config)# int s0/0/1

R2(config-rtr)# ipv6 rip Test2 default-information originate
```

Step 3: Verify the routing configuration.

a. View the IPv6 routing table on R2.

```
R2# show ipv6 route
IPv6 Routing Table - 10 entries
Codes: C - Connected, L - Local, S - Static, R - RIP, B - BGP
       U - Per-user Static route, M - MIPv6
       I1 - ISIS L1, I2 - ISIS L2, IA - ISIS interarea, IS - ISIS summary
       O - OSPF intra, OI - OSPF inter, OE1 - OSPF ext 1, OE2 - OSPF ext 2
       ON1 - OSPF NSSA ext 1, ON2 - OSPF NSSA ext 2
       D - EIGRP, EX - EIGRP external
S    ::/64 [1/0]
       via
R    2001:DB8:ACAD:A::/64 [120/2]
       via FE80::1, Serial0/0/0
C    2001:DB8:ACAD:B::/64 [0/0]
       via ::, GigabitEthernet0/1
```

```
L    2001:DB8:ACAD:B::2/128 [0/0]

        via ::, GigabitEthernet0/1

R    2001:DB8:ACAD:C::/64 [120/2]

        via FE80::3, Serial0/0/1

C    2001:DB8:ACAD:12::/64 [0/0]

        via ::, Serial0/0/0

L    2001:DB8:ACAD:12::2/128 [0/0]

        via ::, Serial0/0/0

C    2001:DB8:ACAD:23::/64 [0/0]

        via ::, Serial0/0/1

L    2001:DB8:ACAD:23::2/128 [0/0]

        via ::, Serial0/0/1

L    FF00::/8 [0/0]

        via ::, Null0
```

How can you tell from the routing table that R2 has a pathway for Internet traffic?

_____

b. View the routing tables on R1 and R3.

How is the pathway for Internet traffic provided in their routing tables?

_____

Step 4: **Verify connectivity.**

Simulate sending traffic to the Internet by pinging from PC-A and PC-C to 2001:DB8:ACAD:B::B/64.

Were the pings successful? _____

Reflection

1. Why would you turn off automatic summarization for RIPv2?

    _____

2. In both scenarios, how did R1 and R3 learn the pathway to the Internet? _____

    _____

3. How are configuring RIPv2 and RIPng different?

    _____

    _____

Router Interface Summary Table

Router Interface Summary				
Router Model	**Ethernet Interface #1**	**Ethernet Interface #2**	**Serial Interface #1**	**Serial Interface #2**
1800	Fast Ethernet 0/0 (F0/0)	Fast Ethernet 0/1 (F0/1)	Serial 0/0/0 (S0/0/0)	Serial 0/0/1 (S0/0/1)
1900	Gigabit Ethernet 0/0 (G0/0)	Gigabit Ethernet 0/1 (G0/1)	Serial 0/0/0 (S0/0/0)	Serial 0/0/1 (S0/0/1)
2801	Fast Ethernet 0/0 (F0/0)	Fast Ethernet 0/1 (F0/1)	Serial 0/1/0 (S0/1/0)	Serial 0/1/1 (S0/1/1)
2811	Fast Ethernet 0/0 (F0/0)	Fast Ethernet 0/1 (F0/1)	Serial 0/0/0 (S0/0/0)	Serial 0/0/1 (S0/0/1)
2900	Gigabit Ethernet 0/0 (G0/0)	Gigabit Ethernet 0/1 (G0/1)	Serial 0/0/0 (S0/0/0)	Serial 0/0/1 (S0/0/1)
Note: To find out how the router is configured, look at the interfaces to identify the type of router and how many interfaces the router has. There is no way to effectively list all the combinations of configurations for each router class. This table includes identifiers for the possible combinations of Ethernet and Serial interfaces in the device. The table does not include any other type of interface, even though a specific router may contain one. An example of this might be an ISDN BRI interface. The string in parenthesis is the legal abbreviation that can be used in Cisco IOS commands to represent the interface.				

7.6.1.1 Class Activity – IPv6 - Details, Details...

Objectives

Analyze a routing table to determine the route source, administrative distance, and metric for a given route to include IPv4/IPv6.

Scenario

After studying the concepts presented in this chapter concerning IPv6, you should be able to read a routing table easily and interpret the IPv6 routing information listed within it.

With a partner, use the IPv6 routing table diagram and the .pdf provided with this activity. Record your answers to the Reflection questions. Then compare your answers with, at least, one other group from the class.

Required Resources

- Routing Table Diagram (as shown below)
- Two PCs or bring your own devices (BYODs): one PC or BYOD will display the Routing Table Diagram for your group to access while recording answers to the Reflection questions on the other PC or BYOD.

Routing Table Diagram

```
R3# show ipv6 route

IPv6 Routing Table - default - 8 entries

Codes: C - Connected, L - Local, S - Static, U - Per-user Static route

       B - BGP, R - RIP, I1 - ISIS L1, I2 - ISIS L2

       IA - ISIS interarea, IS - ISIS summary, D - EIGRP, EX - EIGRP external

       ND - ND Default, NDp - ND Prefix, DCE - Destination, NDr - Redirect

       O - OSPF Intra, OI - OSPF Inter, OE1 - OSPF ext 1, OE2 - OSPF ext 2

       ON1 - OSPF NSSA ext 1, ON2 - OSPF NSSA ext 2

R   2001:DB8:CAFE:1::/64 [120/3]

     via FE80::FE99:47FF:FE71:78A0, Serial0/0/1

R   2001:DB8:CAFE:2::/64 [120/2]

     via FE80::FE99:47FF:FE71:78A0, Serial0/0/1

C   2001:DB8:CAFE:3::/64 [0/0]

     via GigabitEthernet0/0, directly connected

L   2001:DB8:CAFE:3::1/128 [0/0]

     via GigabitEthernet0/0, receive

(output omitted)
```

Reflection

1. How many different IPv6 networks are shown on the routing table diagram? List them in the table provided below.

Routing Table IPv6 Networks

2. The 2001:DB8:CAFE:3:: route is listed twice on the routing table, once with a /64 and once with a /128. What is the significance of this dual network entry?

3. How many routes in this table are RIP routes? What type of RIP routes are listed: RIP, RIPv2, or RIPng?

4. Use the first RIP route, as listed on the routing table, as a reference. What is the administrative distance of this route? What is the cost? What is the significance of these two values?

5. Use the second RIP route, as referenced by the routing table diagram. How many hops would it take to get to the 2001:DB8:CAFE:2::/64 network? What would happen to this routing table entry if the cost for this route exceeded 15 hops?

6. You are designing an IPv6 addressing scheme to add another router to your network's physical topology. Use the /64 prefix for this addressing scheme and an IPv6 network base of 2001:DB8:CAFF:2::/64,. What would be the next, numerical network assignment you could use if the first three hextets remained the same? Justify your answer.

Chapter 8 — Single-Area OSPF

8.0.1.2 Class Activity – Can Submarines Swim?

Objectives

Explain the process by which link-state routers learn about other networks.

Scenario

Edsger Wybe Dijkstra was a famous computer programmer and theoretical physicist. One of his most famous quotes was: "The question of whether computers can think is like the question of whether submarines can swim." Dijkstra's work has been applied, among other things, to routing protocols. He created the Shortest Path First (SPF) algorithm for network routing.

- Visit the Association for Computing Machinery's (ACM) website at http://amturing.acm.org/award_winners/dijkstra_1053701.cfm. Read the article about the life of Dijkstra. List five facts from the article you found interesting about him and his work.

- Next, view Dijkstra's animation of how to find the shortest path first located at http://upload.wikimedia.org/wikipedia/commons/5/57/Dijkstra_Animation.gif. While viewing the animation, pay close attention to what is occurring in it. Note three observations about the animation.

- Lastly, view the graphic located at http://upload.wikimedia.org/wikipedia/commons/3/37/Ricerca_operativa_percorso_minimo_01.gif. Take a few moments to view the visual and notate three observations you have made about the visual. (Note: Use a web translator if you do not know the Italian words "Casa" and "Ufficio".)

Now, open the PDF provided with this activity and answer the reflection questions. Save your work.

Get together with two of your classmates to compare your answers.

Resources

- Internet connection
- Internet browser

Reflection

1. List five facts you found interesting about Edsger Wybe Dijkstra's life.

_____

2. List three observations about the animation found at http://upload.wikimedia.org/wikipedia/commons/5/57/Dijkstra_Animation.gif.

_____

3. List three observations about the visual shown at http://commons.wikimedia.org/wiki/File:Ricerca_operativa_ percorso_minimo_01.gif.

4. Distance vector routing protocols basically depend on number of hops to find the best route from source to destination. If you apply the information you learned from this introductory activity to routing, would hops be the main factor in finding the best path from source to destination? If compared to network communication, could it possibly be better to find the best path using a different metric than hop count? Justify your answer.

8.2.4.5 Lab – Configuring Basic Single-Area OSPFv2

Topology

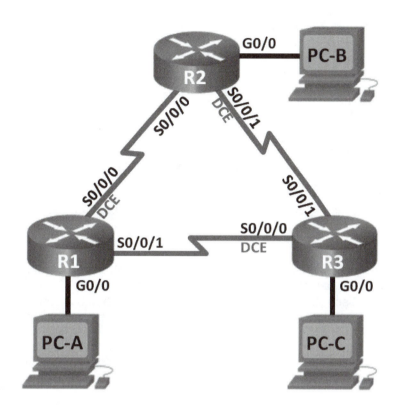

Addressing Table

Device	Interface	IP Address	Subnet Mask	Default Gateway
R1	G0/0	192.168.1.1	255.255.255.0	N/A
	S0/0/0 (DCE)	192.168.12.1	255.255.255.252	N/A
	S0/0/1	192.168.13.1	255.255.255.252	N/A
R2	G0/0	192.168.2.1	255.255.255.0	N/A
	S0/0/0	192.168.12.2	255.255.255.252	N/A
	S0/0/1 (DCE)	192.168.23.1	255.255.255.252	N/A
R3	G0/0	192.168.3.1	255.255.255.0	N/A
	S0/0/0 (DCE)	192.168.13.2	255.255.255.252	N/A
	S0/0/1	192.168.23.2	255.255.255.252	N/A
PC-A	NIC	192.168.1.3	255.255.255.0	192.168.1.1
PC-B	NIC	192.168.2.3	255.255.255.0	192.168.2.1
PC-C	NIC	192.168.3.3	255.255.255.0	192.168.3.1

Objectives

Part 1: Build the Network and Configure Basic Device Settings

Part 2: Configure and Verify OSPF Routing

Part 3: Change Router ID Assignments

Part 4: Configure OSPF Passive Interfaces

Part 5: Change OSPF Metrics

Background / Scenario

Open Shortest Path First (OSPF) is a link-state routing protocol for IP networks. OSPFv2 is defined for IPv4 networks, and OSPFv3 is defined for IPv6 networks. OSPF detects changes in the topology, such as link failures, and converges on a new loop-free routing structure very quickly. It computes each route using Dijkstra's algorithm, a shortest path first algorithm.

In this lab, you will configure the network topology with OSPFv2 routing, change the router ID assignments, configure passive interfaces, adjust OSPF metrics, and use a number of CLI commands to display and verify OSPF routing information.

Note: The routers used with CCNA hands-on labs are Cisco 1941 Integrated Services Routers (ISRs) with Cisco IOS Release 15.2(4)M3 (universalk9 image). Other routers and Cisco IOS versions can be used. Depending on the model and Cisco IOS version, the commands available and output produced might vary from what is shown in the labs. Refer to the Router Interface Summary Table at the end of this lab for the correct interface identifiers.

Note: Make sure that the routers have been erased and have no startup configurations. If you are unsure, contact your instructor.

Required Resources

- 3 Routers (Cisco 1941 with Cisco IOS Release 15.2(4)M3 universal image or comparable)
- 3 PCs (Windows 7, Vista, or XP with terminal emulation program, such as Tera Term)
- Console cables to configure the Cisco IOS devices via the console ports
- Ethernet and serial cables as shown in the topology

Part 1:   Build the Network and Configure Basic Device Settings

In Part 1, you set up the network topology and configure basic settings on the PC hosts and routers.

Step 1:   Cable the network as shown in the topology.

Step 2:   Initialize and reload the routers as necessary.

Step 3:   Configure basic settings for each router.

a.   Disable DNS lookup.

b.   Configure device name as shown in the topology.

c. Assign **class** as the privileged EXEC password.

d. Assign **cisco** as the console and vty passwords.

e. Configure a message of the day (MOTD) banner to warn users that unauthorized access is prohibited.

f. Configure **logging synchronous** for the console line.

g. Configure the IP address listed in the Addressing Table for all interfaces.

h. Set the clock rate for all DCE serial interfaces at **128000**.

i. Copy the running configuration to the startup configuration.

Step 4: Configure PC hosts.

Step 5: Test connectivity.

The routers should be able to ping one another, and each PC should be able to ping its default gateway. The PCs are unable to ping other PCs until OSPF routing is configured. Verify and troubleshoot if necessary.

Part 2: Configure and Verify OSPF Routing

In Part 2, you will configure OSPFv2 routing on all routers in the network and then verify that routing tables are updated correctly. After OSPF has been verified, you will configure OSPF authentication on the links for added security.

Step 1: Configure OSPF on R1.

a. Use the **router ospf** command in global configuration mode to enable OSPF on R1.

```
R1(config)# router ospf 1
```

Note: The OSPF process id is kept locally and has no meaning to other routers on the network.

b. Configure the **network** statements for the networks on R1. Use an area ID of 0.

```
R1(config-router)# network 192.168.1.0 0.0.0.255 area 0
R1(config-router)# network 192.168.12.0 0.0.0.3 area 0
R1(config-router)# network 192.168.13.0 0.0.0.3 area 0
```

Step 2: Configure OSPF on R2 and R3.

Use the **router ospf** command and add the **network** statements for the networks on R2 and R3. Neighbor adjacency messages display on R1 when OSPF routing is configured on R2 and R3.

```
R1#

00:22:29: %OSPF-5-ADJCHG: Process 1, Nbr 192.168.23.1 on Serial0/0/0 from LOADING to
FULL, Loading Done

R1#

00:23:14: %OSPF-5-ADJCHG: Process 1, Nbr 192.168.23.2 on Serial0/0/1 from LOADING to
FULL, Loading Done

R1#
```

Step 3: **Verify OSPF neighbors and routing information.**

a. Issue the **show ip ospf neighbor** command to verify that each router lists the other routers in the network as neighbors.

```
R1# show ip ospf neighbor


Neighbor ID      Pri   State           Dead Time    Address          Interface
192.168.23.2      0    FULL/ -         00:00:33     192.168.13.2     Serial0/0/1
192.168.23.1      0    FULL/ -         00:00:30     192.168.12.2     Serial0/0/0
```

b. Issue the **show ip route** command to verify that all networks display in the routing table on all routers.

```
R1# show ip route
Codes: L - local, C - connected, S - static, R - RIP, M - mobile, B - BGP
       D - EIGRP, EX - EIGRP external, O - OSPF, IA - OSPF inter area
       N1 - OSPF NSSA external type 1, N2 - OSPF NSSA external type 2
       E1 - OSPF external type 1, E2 - OSPF external type 2, E - EGP
       i - IS-IS, L1 - IS-IS level-1, L2 - IS-IS level-2, ia - IS-IS inter area
       * - candidate default, U - per-user static route, o - ODR
       P - periodic downloaded static route


Gateway of last resort is not set


      192.168.1.0/24 is variably subnetted, 2 subnets, 2 masks
C        192.168.1.0/24 is directly connected, GigabitEthernet0/0
L        192.168.1.1/32 is directly connected, GigabitEthernet0/0
O        192.168.2.0/24 [110/65] via 192.168.12.2, 00:32:33, Serial0/0/0
O        192.168.3.0/24 [110/65] via 192.168.13.2, 00:31:48, Serial0/0/1
      192.168.12.0/24 is variably subnetted, 2 subnets, 2 masks
C        192.168.12.0/30 is directly connected, Serial0/0/0
L        192.168.12.1/32 is directly connected, Serial0/0/0
      192.168.13.0/24 is variably subnetted, 2 subnets, 2 masks
C        192.168.13.0/30 is directly connected, Serial0/0/1
L        192.168.13.1/32 is directly connected, Serial0/0/1
      192.168.23.0/30 is subnetted, 1 subnets
O        192.168.23.0/30 [110/128] via 192.168.12.2, 00:31:38, Serial0/0/0
                         [110/128] via 192.168.13.2, 00:31:38, Serial0/0/1
```

What command would you use to only see the OSPF routes in the routing table?

Step 4: Verify OSPF protocol settings.

The **show ip protocols** command is a quick way to verify vital OSPF configuration information. This information includes the OSPF process ID, the router ID, networks the router is advertising, the neighbors the router is receiving updates from, and the default administrative distance, which is 110 for OSPF.

```
R1# show ip protocols
*** IP Routing is NSF aware ***


Routing Protocol is "ospf 1"
  Outgoing update filter list for all interfaces is not set
  Incoming update filter list for all interfaces is not set
  Router ID 192.168.13.1
  Number of areas in this router is 1. 1 normal 0 stub 0 nssa
  Maximum path: 4
  Routing for Networks:
    192.168.1.0 0.0.0.255 area 0
    192.168.12.0 0.0.0.3 area 0
    192.168.13.0 0.0.0.3 area 0
  Routing Information Sources:
    Gateway         Distance      Last Update
    192.168.23.2         110      00:19:16
    192.168.23.1         110      00:20:03
  Distance: (default is 110)
```

Step 5: Verify OSPF process information.

Use the **show ip ospf command** to examine the OSPF process ID and router ID. This command displays the OSPF area information, as well as the last time the SPF algorithm was calculated.

```
R1# show ip ospf
Routing Process "ospf 1" with ID 192.168.13.1
Start time: 00:20:23.260, Time elapsed: 00:25:08.296
Supports only single TOS(TOS0) routes
Supports opaque LSA
Supports Link-local Signaling (LLS)
Supports area transit capability
Supports NSSA (compatible with RFC 3101)
Event-log enabled, Maximum number of events: 1000, Mode: cyclic
Router is not originating router-LSAs with maximum metric
Initial SPF schedule delay 5000 msecs
Minimum hold time between two consecutive SPFs 10000 msecs
Maximum wait time between two consecutive SPFs 10000 msecs
Incremental-SPF disabled
Minimum LSA interval 5 secs
```

```
        Minimum LSA arrival 1000 msecs

        LSA group pacing timer 240 secs

        Interface flood pacing timer 33 msecs

        Retransmission pacing timer 66 msecs

        Number of external LSA 0. Checksum Sum 0x000000

        Number of opaque AS LSA 0. Checksum Sum 0x000000

        Number of DCbitless external and opaque AS LSA 0

        Number of DoNotAge external and opaque AS LSA 0

        Number of areas in this router is 1. 1 normal 0 stub 0 nssa

        Number of areas transit capable is 0

        External flood list length 0

        IETF NSF helper support enabled

        Cisco NSF helper support enabled

        Reference bandwidth unit is 100 mbps

            Area BACKBONE(0)

                Number of interfaces in this area is 3

                Area has no authentication

                SPF algorithm last executed 00:22:53.756 ago

                SPF algorithm executed 7 times

                Area ranges are

                Number of LSA 3. Checksum Sum 0x019A61

                Number of opaque link LSA 0. Checksum Sum 0x000000

                Number of DCbitless LSA 0

                Number of indication LSA 0

                Number of DoNotAge LSA 0

                Flood list length 0
```

Step 6: Verify OSPF interface settings.

a. Issue the **show ip ospf interface brief** command to display a summary of OSPF-enabled interfaces.

```
R1# show ip ospf interface brief
Interface     PID   Area          IP Address/Mask    Cost   State Nbrs F/C
Se0/0/1       1     0             192.168.13.1/30    64     P2P   1/1
Se0/0/0       1     0             192.168.12.1/30    64     P2P   1/1
Gi0/0         1     0             192.168.1.1/24     1      DR    0/0
```

b. For a more detailed list of every OSPF-enabled interface, issue the **show ip ospf interface** command.

```
R1# show ip ospf interface
Serial0/0/1 is up, line protocol is up
  Internet Address 192.168.13.1/30, Area 0, Attached via Network Statement
  Process ID 1, Router ID 192.168.13.1, Network Type POINT_TO_POINT, Cost: 64
  Topology-MTID    Cost    Disabled    Shutdown    Topology Name
        0           64        no          no          Base
```

Transmit Delay is 1 sec, State POINT_TO_POINT

Timer intervals configured, Hello 10, Dead 40, Wait 40, Retransmit 5

  oob-resync timeout 40

  Hello due in 00:00:01

Supports Link-local Signaling (LLS)

Cisco NSF helper support enabled

IETF NSF helper support enabled

Index 3/3, flood queue length 0

Next 0x0(0)/0x0(0)

Last flood scan length is 1, maximum is 1

Last flood scan time is 0 msec, maximum is 0 msec

Neighbor Count is 1, Adjacent neighbor count is 1

  Adjacent with neighbor 192.168.23.2

Suppress hello for 0 neighbor(s)

Serial0/0/0 is up, line protocol is up

  Internet Address 192.168.12.1/30, Area 0, Attached via Network Statement

  Process ID 1, Router ID 192.168.13.1, Network Type POINT_TO_POINT, Cost: 64

Topology-MTID	Cost	Disabled	Shutdown	Topology Name
0	64	no	no	Base

  Transmit Delay is 1 sec, State POINT_TO_POINT

  Timer intervals configured, Hello 10, Dead 40, Wait 40, Retransmit 5

    oob-resync timeout 40

    Hello due in 00:00:03

  Supports Link-local Signaling (LLS)

  Cisco NSF helper support enabled

  IETF NSF helper support enabled

  Index 2/2, flood queue length 0

  Next 0x0(0)/0x0(0)

  Last flood scan length is 1, maximum is 1

  Last flood scan time is 0 msec, maximum is 0 msec

  Neighbor Count is 1, Adjacent neighbor count is 1

    Adjacent with neighbor 192.168.23.1

  Suppress hello for 0 neighbor(s)

GigabitEthernet0/0 is up, line protocol is up

  Internet Address 192.168.1.1/24, Area 0, Attached via Network Statement

  Process ID 1, Router ID 192.168.13.1, Network Type BROADCAST, Cost: 1

Topology-MTID	Cost	Disabled	Shutdown	Topology Name
0	1	no	no	Base

  Transmit Delay is 1 sec, State DR, Priority 1

  Designated Router (ID) 192.168.13.1, Interface address 192.168.1.1

  No backup designated router on this network

```
    Timer intervals configured, Hello 10, Dead 40, Wait 40, Retransmit 5
      oob-resync timeout 40
      Hello due in 00:00:01
    Supports Link-local Signaling (LLS)
    Cisco NSF helper support enabled
    IETF NSF helper support enabled
    Index 1/1, flood queue length 0
    Next 0x0(0)/0x0(0)
    Last flood scan length is 0, maximum is 0
    Last flood scan time is 0 msec, maximum is 0 msec
    Neighbor Count is 0, Adjacent neighbor count is 0
    Suppress hello for 0 neighbor(s)
```

Step 7: **Verify end-to-end connectivity.**

Each PC should be able to ping the other PCs in the topology. Verify and troubleshoot if necessary.

Note: It may be necessary to disable the PC firewall to ping between PCs.

Part 3: **Change Router ID Assignments**

The OSPF router ID is used to uniquely identify the router in the OSPF routing domain. Cisco routers derive the router ID in one of three ways and with the following precedence:

1) IP address configured with the OSPF **router-id** command, if present

2) Highest IP address of any of the router's loopback addresses, if present

3) Highest active IP address on any of the router's physical interfaces

Because no router IDs or loopback interfaces have been configured on the three routers, the router ID for each router is determined by the highest IP address of any active interface.

In Part 3, you will change the OSPF router ID assignment using loopback addresses. You will also use the **router-id** command to change the router ID.

Step 1: **Change router IDs using loopback addresses.**

a. Assign an IP address to loopback 0 on R1.

```
R1(config)# interface lo0
R1(config-if)# ip address 1.1.1.1 255.255.255.255
R1(config-if)# end
```

b. Assign IP addresses to Loopback 0 on R2 and R3. Use IP address 2.2.2.2/32 for R2 and 3.3.3.3/32 for R3.

c. Save the running configuration to the startup configuration on all three routers.

d. You must reload the routers in order to reset the router ID to the loopback address. Issue the **reload** command on all three routers. Press Enter to confirm the reload.

e. After the router completes the reload process, issue the **show ip protocols** command to view the new router ID.

```
R1# show ip protocols
*** IP Routing is NSF aware ***


Routing Protocol is "ospf 1"
  Outgoing update filter list for all interfaces is not set
  Incoming update filter list for all interfaces is not set
  Router ID 1.1.1.1
  Number of areas in this router is 1. 1 normal 0 stub 0 nssa
  Maximum path: 4
  Routing for Networks:
    192.168.1.0 0.0.0.255 area 0
    192.168.12.0 0.0.0.3 area 0
    192.168.13.0 0.0.0.3 area 0
  Routing Information Sources:
    Gateway         Distance      Last Update
    3.3.3.3              110      00:01:00
    2.2.2.2              110      00:01:14
  Distance: (default is 110)
```

f. Issue the **show ip ospf neighbor** command to display the router ID changes for the neighboring routers.

```
R1# show ip ospf neighbor


Neighbor ID     Pri   State          Dead Time   Address         Interface
3.3.3.3           0   FULL/  -       00:00:35    192.168.13.2    Serial0/0/1
2.2.2.2           0   FULL/  -       00:00:32    192.168.12.2    Serial0/0/0
R1#
```

Step 1: Change the router ID on R1 using the router-id command.

The preferred method for setting the router ID is with the **router-id** command.

a. Issue the **router-id 11.11.11.11** command on R1 to reassign the router ID. Notice the informational message that appears when issuing the **router-id** command.

```
R1(config)# router ospf 1
R1(config-router)# router-id 11.11.11.11
Reload or use "clear ip ospf process" command, for this to take effect
R1(config)# end
```

b. You will receive an informational message telling you that you must either reload the router or use the **clear ip ospf process** command for the change to take effect. Issue the **clear ip ospf process** command on all three routers. Type **yes** to reply to the reset verification message, and press ENTER.

c. Set the router ID for R2 to **22.22.22.22** and the router ID for R3 to **33.33.33.33**. Then use **clear ip ospf process** command to reset ospf routing process.

d. Issue the **show ip protocols** command to verify that the router ID changed on R1.

```
R1# show ip protocols

*** IP Routing is NSF aware ***


Routing Protocol is "ospf 1"
  Outgoing update filter list for all interfaces is not set
  Incoming update filter list for all interfaces is not set
  Router ID 11.11.11.11
  Number of areas in this router is 1. 1 normal 0 stub 0 nssa
  Maximum path: 4
  Routing for Networks:
    192.168.1.0 0.0.0.255 area 0
    192.168.12.0 0.0.0.3 area 0
    192.168.13.0 0.0.0.3 area 0
  Passive Interface(s):
    GigabitEthernet0/1
  Routing Information Sources:
    Gateway          Distance      Last Update
    33.33.33.33           110      00:00:19
    22.22.22.22           110      00:00:31
    3.3.3.3               110      00:00:41
    2.2.2.2               110      00:00:41
  Distance: (default is 110)
```

e. Issue the **show ip ospf neighbor** command on R1 to verify that new router ID for R2 and R3 is listed.

```
R1# show ip ospf neighbor


Neighbor ID      Pri    State        Dead Time    Address        Interface
33.33.33.33        0    FULL/  -     00:00:36     192.168.13.2   Serial0/0/1
22.22.22.22        0    FULL/  -     00:00:32     192.168.12.2   Serial0/0/0
```

Part 4: Configure OSPF Passive Interfaces

The **passive-interface** command prevents routing updates from being sent through the specified router interface. This is commonly done to reduce traffic on the LANs as they do not need to receive dynamic routing protocol communication. In Part 4, you will use the **passive-interface** command to configure a single interface as passive. You will also configure OSPF so that all interfaces on the router are passive by default, and then enable OSPF routing advertisements on selected interfaces.

Step 1: Configure a passive interface.

a. Issue the **show ip ospf interface g0/0** command on R1. Notice the timer indicating when the next Hello packet is expected. Hello packets are sent every 10 seconds and are used between OSPF routers to verify that their neighbors are up.

```
R1# show ip ospf interface g0/0
GigabitEthernet0/0 is up, line protocol is up
  Internet Address 192.168.1.1/24, Area 0, Attached via Network Statement
  Process ID 1, Router ID 11.11.11.11, Network Type BROADCAST, Cost: 1
  Topology-MTID    Cost    Disabled    Shutdown      Topology Name
        0           1         no          no             Base
  Transmit Delay is 1 sec, State DR, Priority 1
  Designated Router (ID) 11.11.11.11, Interface address 192.168.1.1
  No backup designated router on this network
  Timer intervals configured, Hello 10, Dead 40, Wait 40, Retransmit 5
    oob-resync timeout 40
    Hello due in 00:00:02
  Supports Link-local Signaling (LLS)
  Cisco NSF helper support enabled
  IETF NSF helper support enabled
  Index 1/1, flood queue length 0
  Next 0x0(0)/0x0(0)
  Last flood scan length is 0, maximum is 0
  Last flood scan time is 0 msec, maximum is 0 msec
  Neighbor Count is 0, Adjacent neighbor count is 0
  Suppress hello for 0 neighbor(s)
```

b. Issue the **passive-interface** command to change the G0/0 interface on R1 to passive.

```
R1(config)# router ospf 1
R1(config-router)# passive-interface g0/0
```

c. Re-issue the **show ip ospf interface g0/0** command to verify that G0/0 is now passive.

```
R1# show ip ospf interface g0/0
GigabitEthernet0/0 is up, line protocol is up
  Internet Address 192.168.1.1/24, Area 0, Attached via Network Statement
  Process ID 1, Router ID 11.11.11.11, Network Type BROADCAST, Cost: 1
  Topology-MTID    Cost    Disabled    Shutdown      Topology Name
        0           1         no          no             Base
  Transmit Delay is 1 sec, State DR, Priority 1
  Designated Router (ID) 11.11.11.11, Interface address 192.168.1.1
  No backup designated router on this network
  Timer intervals configured, Hello 10, Dead 40, Wait 40, Retransmit 5
    oob-resync timeout 40
    No Hellos (Passive interface)
  Supports Link-local Signaling (LLS)
  Cisco NSF helper support enabled
  IETF NSF helper support enabled
```

```
Index 1/1, flood queue length 0

Next 0x0(0)/0x0(0)

Last flood scan length is 0, maximum is 0

Last flood scan time is 0 msec, maximum is 0 msec

Neighbor Count is 0, Adjacent neighbor count is 0

Suppress hello for 0 neighbor(s)
```

d.   Issue the **show ip route** command on R2 and R3 to verify that a route to the 192.168.1.0/24 network is still available.

```
R2# show ip route

Codes: L - local, C - connected, S - static, R - RIP, M - mobile, B - BGP

       D - EIGRP, EX - EIGRP external, O - OSPF, IA - OSPF inter area

       N1 - OSPF NSSA external type 1, N2 - OSPF NSSA external type 2

       E1 - OSPF external type 1, E2 - OSPF external type 2

       i - IS-IS, su - IS-IS summary, L1 - IS-IS level-1, L2 - IS-IS level-2

       ia - IS-IS inter area, * - candidate default, U - per-user static route

       o - ODR, P - periodic downloaded static route, H - NHRP, l - LISP

       + - replicated route, % - next hop override


Gateway of last resort is not set


      2.0.0.0/32 is subnetted, 1 subnets

C        2.2.2.2 is directly connected, Loopback0

O     192.168.1.0/24 [110/65] via 192.168.12.1, 00:58:32, Serial0/0/0

      192.168.2.0/24 is variably subnetted, 2 subnets, 2 masks

C        192.168.2.0/24 is directly connected, GigabitEthernet0/0

L        192.168.2.1/32 is directly connected, GigabitEthernet0/0

O     192.168.3.0/24 [110/65] via 192.168.23.2, 00:58:19, Serial0/0/1

      192.168.12.0/24 is variably subnetted, 2 subnets, 2 masks

C        192.168.12.0/30 is directly connected, Serial0/0/0

L        192.168.12.2/32 is directly connected, Serial0/0/0

      192.168.13.0/30 is subnetted, 1 subnets

O        192.168.13.0 [110/128] via 192.168.23.2, 00:58:19, Serial0/0/1

                     [110/128] via 192.168.12.1, 00:58:32, Serial0/0/0

      192.168.23.0/24 is variably subnetted, 2 subnets, 2 masks

C        192.168.23.0/30 is directly connected, Serial0/0/1

L        192.168.23.1/32 is directly connected, Serial0/0/1
```

Step 2:   **Set passive interface as the default on a router.**

a.   Issue the **show ip ospf neighbor** command on R1 to verify that R2 is listed as an OSPF neighbor.

```
R1# show ip ospf neighbor
```

Neighbor ID	Pri	State	Dead Time	Address	Interface
33.33.33.33	0	FULL/ -	00:00:31	192.168.13.2	Serial0/0/1
22.22.22.22	0	FULL/ -	00:00:32	192.168.12.2	Serial0/0/0

b. Issue the **passive-interface default** command on R2 to set the default for all OSPF interfaces as passive.

```
R2(config)# router ospf 1
```

```
R2(config-router)# passive-interface default
```

```
R2(config-router)#
```

```
*Apr  3 00:03:00.979: %OSPF-5-ADJCHG: Process 1, Nbr 11.11.11.11 on Serial0/0/0 from
FULL to DOWN, Neighbor Down: Interface down or detached
```

```
*Apr  3 00:03:00.979: %OSPF-5-ADJCHG: Process 1, Nbr 33.33.33.33 on Serial0/0/1 from
FULL to DOWN, Neighbor Down: Interface down or detached
```

c. Re-issue the **show ip ospf neighbor** command on R1. After the dead timer expires, R2 will no longer be listed as an OSPF neighbor.

```
R1# show ip ospf neighbor
```

Neighbor ID	Pri	State	Dead Time	Address	Interface
33.33.33.33	0	FULL/ -	00:00:34	192.168.13.2	Serial0/0/1

d. Issue the **show ip ospf interface S0/0/0** command on R2 to view the OSPF status of interface S0/0/0.

```
R2# show ip ospf interface s0/0/0
Serial0/0/0 is up, line protocol is up
  Internet Address 192.168.12.2/30, Area 0, Attached via Network Statement
  Process ID 1, Router ID 22.22.22.22, Network Type POINT_TO_POINT, Cost: 64
  Topology-MTID    Cost    Disabled    Shutdown      Topology Name
        0           64        no          no             Base
  Transmit Delay is 1 sec, State POINT_TO_POINT
  Timer intervals configured, Hello 10, Dead 40, Wait 40, Retransmit 5
    oob-resync timeout 40
    No Hellos (Passive interface)
  Supports Link-local Signaling (LLS)
  Cisco NSF helper support enabled
  IETF NSF helper support enabled
  Index 2/2, flood queue length 0
  Next 0x0(0)/0x0(0)
  Last flood scan length is 0, maximum is 0
  Last flood scan time is 0 msec, maximum is 0 msec
  Neighbor Count is 0, Adjacent neighbor count is 0
  Suppress hello for 0 neighbor(s)
```

e. If all interfaces on R2 are passive, then no routing information is being advertised. In this case, R1 and R3 should no longer have a route to the 192.168.2.0/24 network. You can verify this by using the **show ip route** command.

f.   On R2, issue the **no passive-interface** command so the router will send and receive OSPF routing up-
     dates. After entering this command, you will see an informational message that a neighbor adjacency has
     been established with R1.

```
R2(config)# router ospf 1

R2(config-router)# no passive-interface s0/0/0

R2(config-router)#

*Apr  3 00:18:03.463: %OSPF-5-ADJCHG: Process 1, Nbr 11.11.11.11 on Serial0/0/0 from
LOADING to FULL, Loading Done
```

g.   Re-issue the **show ip route** and **show ipv6 ospf neighbor** commands on R1 and R3, and look for a
     route to the 192.168.2.0/24 network.

     What interface is R3 using to route to the 192.168.2.0/24 network? _____

     What is the accumulated cost metric for the 192.168.2.0/24 network on R3? _____

     Does R2 show up as an OSPF neighbor on R1? _____

     Does R2 show up as an OSPF neighbor on R3? _____

     What does this information tell you?

     _____

     _____

     _____

     _____

h.   Change interface S0/0/1 on R2 to allow it to advertise OSPF routes. Record the commands used below.

     _____

     _____

i.   Re-issue the **show ip route** command on R3.

     What interface is R3 using to route to the 192.168.2.0/24 network? _____

     What is the accumulated cost metric for the 192.168.2.0/24 network on R3 now and how is this calcu-
     lated?

     _____

     Is R2 listed as an OSPF neighbor to R3? _____

Part 5:   Change OSPF Metrics

In Part 3, you will change OSPF metrics using the **auto-cost reference-bandwidth** command, the **bandwidth** command, and the **ip ospf cost** command.

Note: All DCE interfaces should have been configured with a clocking rate of 128000 in Part 1.

Step 1:   Change the reference bandwidth on the routers.

The default reference-bandwidth for OSPF is 100Mb/s (Fast Ethernet speed). However, most modern infrastructure devices have links that are faster than 100Mb/s. Because the OSPF cost metric must be an integer, all links with transmission speeds of 100Mb/s or higher have a cost of 1. This results in Fast Ethernet, Gigabit Ethernet and 10G Ethernet interfaces all having the same cost. Therefore, the reference-bandwidth must be changed to a higher value to accommodate networks with links faster that 100Mb/s.

a.   Issue the **show interface** command on R1 to view the default bandwidth setting for the G0/0 interface.

```
R1# show interface g0/0
GigabitEthernet0/0 is up, line protocol is up
  Hardware is CN Gigabit Ethernet, address is c471.fe45.7520 (bia c471.fe45.7520)
  MTU 1500 bytes, BW 1000000 Kbit/sec, DLY 100 usec,
     reliability 255/255, txload 1/255, rxload 1/255
  Encapsulation ARPA, loopback not set
  Keepalive set (10 sec)
  Full Duplex, 100Mbps, media type is RJ45
  output flow-control is unsupported, input flow-control is unsupported
  ARP type: ARPA, ARP Timeout 04:00:00
  Last input never, output 00:17:31, output hang never
  Last clearing of "show interface" counters never
  Input queue: 0/75/0/0 (size/max/drops/flushes); Total output drops: 0
  Queueing strategy: fifo
  Output queue: 0/40 (size/max)
  5 minute input rate 0 bits/sec, 0 packets/sec
  5 minute output rate 0 bits/sec, 0 packets/sec
     0 packets input, 0 bytes, 0 no buffer
     Received 0 broadcasts (0 IP multicasts)
     0 runts, 0 giants, 0 throttles
     0 input errors, 0 CRC, 0 frame, 0 overrun, 0 ignored
     0 watchdog, 0 multicast, 0 pause input
     279 packets output, 89865 bytes, 0 underruns
     0 output errors, 0 collisions, 1 interface resets
     0 unknown protocol drops
     0 babbles, 0 late collision, 0 deferred
     1 lost carrier, 0 no carrier, 0 pause output
     0 output buffer failures, 0 output buffers swapped out
```

Note: The bandwidth setting on G0/0 may differ from what is shown above if the PC host interface can only support Fast Ethernet speed. If the PC host interface is not capable of supporting gigabit speed, then the bandwidth will most likely be displayed as 100000 Kbit/sec.

b. Issue the **show ip route ospf** command on R1 to determine the route to the 192.168.3.0/24 network.

```
R1# show ip route ospf
Codes: L - local, C - connected, S - static, R - RIP, M - mobile, B - BGP
       D - EIGRP, EX - EIGRP external, O - OSPF, IA - OSPF inter area
       N1 - OSPF NSSA external type 1, N2 - OSPF NSSA external type 2
       E1 - OSPF external type 1, E2 - OSPF external type 2
       i - IS-IS, su - IS-IS summary, L1 - IS-IS level-1, L2 - IS-IS level-2
       ia - IS-IS inter area, * - candidate default, U - per-user static route
       o - ODR, P - periodic downloaded static route, H - NHRP, l - LISP
       + - replicated route, % - next hop override


Gateway of last resort is not set


O     192.168.3.0/24 [110/65] via 192.168.13.2, 00:00:57, Serial0/0/1
      192.168.23.0/30 is subnetted, 1 subnets
O        192.168.23.0 [110/128] via 192.168.13.2, 00:00:57, Serial0/0/1
                      [110/128] via 192.168.12.2, 00:01:08, Serial0/0/0
```

Note: The accumulated cost to the 192.168.3.0/24 network from R1 is 65.

c. Issue the **show ip ospf interface** command on R3 to determine the routing cost for G0/0.

```
R3# show ip ospf interface g0/0
GigabitEthernet0/0 is up, line protocol is up
  Internet Address 192.168.3.1/24, Area 0, Attached via Network Statement
  Process ID 1, Router ID 3.3.3.3, Network Type BROADCAST, Cost: 1
  Topology-MTID    Cost    Disabled    Shutdown      Topology Name
        0            1         no          no            Base
  Transmit Delay is 1 sec, State DR, Priority 1
  Designated Router (ID) 192.168.23.2, Interface address 192.168.3.1
  No backup designated router on this network
  Timer intervals configured, Hello 10, Dead 40, Wait 40, Retransmit 5
    oob-resync timeout 40
    Hello due in 00:00:05
  Supports Link-local Signaling (LLS)
  Cisco NSF helper support enabled
  IETF NSF helper support enabled
  Index 1/1, flood queue length 0
  Next 0x0(0)/0x0(0)
  Last flood scan length is 0, maximum is 0
  Last flood scan time is 0 msec, maximum is 0 msec
```

```
    Neighbor Count is 0, Adjacent neighbor count is 0

    Suppress hello for 0 neighbor(s)
```

d. Issue the **show ip ospf interface s0/0/1** command on R1 to view the routing cost for S0/0/1.

```
R1# show ip ospf interface s0/0/1
Serial0/0/1 is up, line protocol is up
    Internet Address 192.168.13.1/30, Area 0, Attached via Network Statement
    Process ID 1, Router ID 1.1.1.1, Network Type POINT_TO_POINT, Cost: 64
    Topology-MTID    Cost    Disabled    Shutdown    Topology Name
         0            64        no          no          Base
    Transmit Delay is 1 sec, State POINT_TO_POINT
    Timer intervals configured, Hello 10, Dead 40, Wait 40, Retransmit 5
      oob-resync timeout 40
      Hello due in 00:00:04
    Supports Link-local Signaling (LLS)
    Cisco NSF helper support enabled
    IETF NSF helper support enabled
    Index 3/3, flood queue length 0
    Next 0x0(0)/0x0(0)
    Last flood scan length is 1, maximum is 1
    Last flood scan time is 0 msec, maximum is 0 msec
    Neighbor Count is 1, Adjacent neighbor count is 1
      Adjacent with neighbor 192.168.23.2
    Suppress hello for 0 neighbor(s)
```

The sum of the costs of these two interfaces is the accumulated cost for the route to the 192.168.3.0/24 network on R3 (1 + 64 = 65), as can be seen in the output from the **show ip route** command.

e. Issue the **auto-cost reference-bandwidth 10000** command on R1 to change the default reference bandwidth setting. With this setting, 10Gb/s interfaces will have a cost of 1, 1 Gb/s interfaces will have a cost of 10, and 100Mb/s interfaces will have a cost of 100.

```
R1(config)# router ospf 1
R1(config-router)# auto-cost reference-bandwidth 10000
% OSPF: Reference bandwidth is changed.
        Please ensure reference bandwidth is consistent across all routers.
```

f. Issue the **auto-cost reference-bandwidth 10000** command on routers R2 and R3.

g. Re-issue the **show ip ospf interface** command to view the new cost of G0/0 on R3, and S0/0/1 on R1.

```
R3# show ip ospf interface g0/0
GigabitEthernet0/0 is up, line protocol is up
    Internet Address 192.168.3.1/24, Area 0, Attached via Network Statement
    Process ID 1, Router ID 3.3.3.3, Network Type BROADCAST, Cost: 10
    Topology-MTID    Cost    Disabled    Shutdown    Topology Name
         0            10        no          no          Base
```

```
      Transmit Delay is 1 sec, State DR, Priority 1
      Designated Router (ID) 192.168.23.2, Interface address 192.168.3.1
      No backup designated router on this network
      Timer intervals configured, Hello 10, Dead 40, Wait 40, Retransmit 5
        oob-resync timeout 40
        Hello due in 00:00:02
      Supports Link-local Signaling (LLS)
      Cisco NSF helper support enabled
      IETF NSF helper support enabled
      Index 1/1, flood queue length 0
      Next 0x0(0)/0x0(0)
      Last flood scan length is 0, maximum is 0
      Last flood scan time is 0 msec, maximum is 0 msec
      Neighbor Count is 0, Adjacent neighbor count is 0
      Suppress hello for 0 neighbor(s)
```

Note: If the device connected to the G0/0 interface does not support Gigabit Ethernet speed, the cost will be different than the output display. For example, the cost will be 100 for Fast Ethernet speed (100Mb/s).

```
R1# show ip ospf interface s0/0/1
Serial0/0/1 is up, line protocol is up
      Internet Address 192.168.13.1/30, Area 0, Attached via Network Statement
      Process ID 1, Router ID 1.1.1.1, Network Type POINT_TO_POINT, Cost: 6476
      Topology-MTID    Cost    Disabled    Shutdown      Topology Name
            0          6476      no          no            Base
      Transmit Delay is 1 sec, State POINT_TO_POINT
      Timer intervals configured, Hello 10, Dead 40, Wait 40, Retransmit 5
        oob-resync timeout 40
        Hello due in 00:00:05
      Supports Link-local Signaling (LLS)
      Cisco NSF helper support enabled
      IETF NSF helper support enabled
      Index 3/3, flood queue length 0
      Next 0x0(0)/0x0(0)
      Last flood scan length is 1, maximum is 1
      Last flood scan time is 0 msec, maximum is 0 msec
      Neighbor Count is 1, Adjacent neighbor count is 1
        Adjacent with neighbor 192.168.23.2
      Suppress hello for 0 neighbor(s)
```

h. Re-issue the **show ip route ospf** command to view the new accumulated cost for the 192.168.3.0/24 route (10 + 6476 = 6486).

Note: If the device connected to the G0/0 interface does not support Gigabit Ethernet speed, the total cost will be different than the output display. For example, the accumulated cost will be 6576 if G0/0 is operating at Fast Ethernet speed (100Mb/s).

```
R1# show ip route ospf
Codes: L - local, C - connected, S - static, R - RIP, M - mobile, B - BGP
       D - EIGRP, EX - EIGRP external, O - OSPF, IA - OSPF inter area
       N1 - OSPF NSSA external type 1, N2 - OSPF NSSA external type 2
       E1 - OSPF external type 1, E2 - OSPF external type 2
       i - IS-IS, su - IS-IS summary, L1 - IS-IS level-1, L2 - IS-IS level-2
       ia - IS-IS inter area, * - candidate default, U - per-user static route
       o - ODR, P - periodic downloaded static route, H - NHRP, l - LISP
       + - replicated route, % - next hop override


Gateway of last resort is not set


O    192.168.2.0/24 [110/6486] via 192.168.12.2, 00:05:40, Serial0/0/0
O    192.168.3.0/24 [110/6486] via 192.168.13.2, 00:01:08, Serial0/0/1
     192.168.23.0/30 is subnetted, 1 subnets
O       192.168.23.0 [110/12952] via 192.168.13.2, 00:05:17, Serial0/0/1
                     [110/12952] via 192.168.12.2, 00:05:17, Serial0/0/
```

Note: Changing the default reference-bandwidth on the routers from 100 to 10,000 in effect changed the accumulated costs of all routes by a factor of 100, but the cost of each interface link and route is now more accurately reflected.

i. To reset the reference-bandwidth back to its default value, issue the **auto-cost reference-bandwidth 100** command on all three routers.

```
R1(config)# router ospf 1
R1(config-router)# auto-cost reference-bandwidth 100
% OSPF: Reference bandwidth is changed.
        Please ensure reference bandwidth is consistent across all routers.
```

Why would you want to change the OSPF default reference-bandwidth?

_____

_____

Step 2: Change the bandwidth for an interface.

On most serial links, the bandwidth metric will default to 1544 Kbits (that of a T1). If this is not the actual speed of the serial link, the bandwidth setting will need to be changed to match the actual speed to allow the route cost to be calculated correctly in OSPF. Use the **bandwidth** command to adjust the bandwidth setting on an interface.

Note: A common misconception is to assume that the **bandwidth** command will change the physical bandwidth, or speed, of the link. The command modifies the bandwidth metric used by OSPF to calculate routing costs, and does not modify the actual bandwidth (speed) of the link.

a. Issue the **show interface s0/0/0** command on R1 to view the current bandwidth setting on S0/0/0. Even though the clock rate, link speed on this interface was set to 128Kb/s, the bandwidth is still showing 1544Kb/s.

```
R1# show interface s0/0/0
Serial0/0/0 is up, line protocol is up
  Hardware is WIC MBRD Serial
  Internet address is 192.168.12.1/30
  MTU 1500 bytes, BW 1544 Kbit/sec, DLY 20000 usec,
     reliability 255/255, txload 1/255, rxload 1/255
  Encapsulation HDLC, loopback not set
  Keepalive set (10 sec)
<Output omitted>
```

b. Issue the **show ip route ospf** command on R1 to view the accumulated cost for the route to network 192.168.23.0/24 using S0/0/0. Note that there are two equal-cost (128) routes to the 192.168.23.0/24 network, one via S0/0/0 and one via S0/0/1.

```
R1# show ip route ospf
Codes: L - local, C - connected, S - static, R - RIP, M - mobile, B - BGP
       D - EIGRP, EX - EIGRP external, O - OSPF, IA - OSPF inter area
       N1 - OSPF NSSA external type 1, N2 - OSPF NSSA external type 2
       E1 - OSPF external type 1, E2 - OSPF external type 2
       i - IS-IS, su - IS-IS summary, L1 - IS-IS level-1, L2 - IS-IS level-2
       ia - IS-IS inter area, * - candidate default, U - per-user static route
       o - ODR, P - periodic downloaded static route, H - NHRP, l - LISP
       + - replicated route, % - next hop override


Gateway of last resort is not set


O     192.168.3.0/24 [110/65] via 192.168.13.2, 00:00:26, Serial0/0/1
      192.168.23.0/30 is subnetted, 1 subnets
O        192.168.23.0 [110/128] via 192.168.13.2, 00:00:26, Serial0/0/1
                      [110/128] via 192.168.12.2, 00:00:42, Serial0/0/0
```

c. Issue the **bandwidth 128** command to set the bandwidth on S0/0/0 to 128Kb/s.

```
R1(config)# interface s0/0/0
R1(config-if)# bandwidth 128
```

d. Re-issue the **show ip route ospf** command. The routing table no longer displays the route to the 192.168.23.0/24 network over the S0/0/0 interface. This is because the best route, the one with the lowest cost, is now via S0/0/1.

```
R1# show ip route ospf
Codes: L - local, C - connected, S - static, R - RIP, M - mobile, B - BGP
       D - EIGRP, EX - EIGRP external, O - OSPF, IA - OSPF inter area
       N1 - OSPF NSSA external type 1, N2 - OSPF NSSA external type 2
```

```
        E1 - OSPF external type 1, E2 - OSPF external type 2

        i - IS-IS, su - IS-IS summary, L1 - IS-IS level-1, L2 - IS-IS level-2

        ia - IS-IS inter area, * - candidate default, U - per-user static route

        o - ODR, P - periodic downloaded static route, H - NHRP, l - LISP

        + - replicated route, % - next hop override


Gateway of last resort is not set


O       192.168.3.0/24 [110/65] via 192.168.13.2, 00:04:51, Serial0/0/1
        192.168.23.0/30 is subnetted, 1 subnets
O           192.168.23.0 [110/128] via 192.168.13.2, 00:04:51, Serial0/0/1
```

e. Issue the **show ip ospf interface brief** command. The cost for S0/0/0 has changed from 64 to 781 which is an accurate cost representation of the link speed.

```
R1# show ip ospf interface brief

Interface    PID   Area        IP Address/Mask      Cost   State Nbrs F/C
Se0/0/1      1     0           192.168.13.1/30      64     P2P   1/1
Se0/0/0      1     0           192.168.12.1/30      781    P2P   1/1
Gi0/0        1     0           192.168.1.1/24       1      DR    0/0
```

f. Change the bandwidth for interface S0/0/1 to the same setting as S0/0/0 on R1.

g. Re-issue the **show ip route ospf** command to view the accumulated cost of both routes to the 192.168.23.0/24 network. Note that there are again two equal-cost (845) routes to the 192.168.23.0/24 network, one via S0/0/0 and one via S0/0/1.

```
R1# show ip route ospf

Codes: L - local, C - connected, S - static, R - RIP, M - mobile, B - BGP
       D - EIGRP, EX - EIGRP external, O - OSPF, IA - OSPF inter area
       N1 - OSPF NSSA external type 1, N2 - OSPF NSSA external type 2
       E1 - OSPF external type 1, E2 - OSPF external type 2
       i - IS-IS, su - IS-IS summary, L1 - IS-IS level-1, L2 - IS-IS level-2
       ia - IS-IS inter area, * - candidate default, U - per-user static route
       o - ODR, P - periodic downloaded static route, H - NHRP, l - LISP
       + - replicated route, % - next hop override


Gateway of last resort is not set


O       192.168.3.0/24 [110/782] via 192.168.13.2, 00:00:09, Serial0/0/1
        192.168.23.0/30 is subnetted, 1 subnets
O           192.168.23.0 [110/845] via 192.168.13.2, 00:00:09, Serial0/0/1
                         [110/845] via 192.168.12.2, 00:00:09, Serial0/0/0
```

Explain how the costs to the 192.168.3.0/24 and 192.168.23.0/30 networks from R1 were calculated.

_____

_____

_____

h. Issue the **show ip route ospf** command on R3. The accumulated cost of the 192.168.1.0/24 is still showing as 65. Unlike the **clock rate** command, the **bandwidth** command needs to be applied on each side of a serial link.

```
R3# show ip route ospf
Codes: L - local, C - connected, S - static, R - RIP, M - mobile, B - BGP
       D - EIGRP, EX - EIGRP external, O - OSPF, IA - OSPF inter area
       N1 - OSPF NSSA external type 1, N2 - OSPF NSSA external type 2
       E1 - OSPF external type 1, E2 - OSPF external type 2
       i - IS-IS, su - IS-IS summary, L1 - IS-IS level-1, L2 - IS-IS level-2
       ia - IS-IS inter area, * - candidate default, U - per-user static route
       o - ODR, P - periodic downloaded static route, H - NHRP, l - LISP
       + - replicated route, % - next hop override


Gateway of last resort is not set

O      192.168.1.0/24 [110/65] via 192.168.13.1, 00:30:58, Serial0/0/0
       192.168.12.0/30 is subnetted, 1 subnets
O         192.168.12.0 [110/128] via 192.168.23.1, 00:30:58, Serial0/0/1
                       [110/128] via 192.168.13.1, 00:30:58, Serial0/0/0
```

i. Issue the **bandwidth 128** command on all remaining serial interfaces in the topology.

What is the new accumulated cost to the 192.168.23.0/24 network on R1? Why?

_____

_____

Step 3: **Change the route cost.**

OSPF uses the bandwidth setting to calculate the cost for a link by default. However, you can override this calculation by manually setting the cost of a link using the **ip ospf cost** command. Like the **bandwidth** command, **the ip ospf cost** command only affects the side of the link where it was applied.

a. Issue the **show ip route ospf** on R1.

```
R1# show ip route ospf
Codes: L - local, C - connected, S - static, R - RIP, M - mobile, B - BGP
```

```
        D - EIGRP, EX - EIGRP external, O - OSPF, IA - OSPF inter area

        N1 - OSPF NSSA external type 1, N2 - OSPF NSSA external type 2

        E1 - OSPF external type 1, E2 - OSPF external type 2

        i - IS-IS, su - IS-IS summary, L1 - IS-IS level-1, L2 - IS-IS level-2

        ia - IS-IS inter area, * - candidate default, U - per-user static route

        o - ODR, P - periodic downloaded static route, H - NHRP, l - LISP

        + - replicated route, % - next hop override


Gateway of last resort is not set


O       192.168.2.0/24 [110/782] via 192.168.12.2, 00:00:26, Serial0/0/0
O       192.168.3.0/24 [110/782] via 192.168.13.2, 00:02:50, Serial0/0/1
        192.168.23.0/30 is subnetted, 1 subnets
O          192.168.23.0 [110/1562] via 192.168.13.2, 00:02:40, Serial0/0/1
                        [110/1562] via 192.168.12.2, 00:02:40, Serial0/0/0
```

b. Apply the **ip ospf cost 1565** command to the S0/0/1 interface on R1. A cost of 1565 is higher than the accumulated cost of the route through R2 which is 1562.

```
R1(config)# int s0/0/1

R1(config-if)# ip ospf cost 1565
```

c. Re-issue the **show ip route ospf** command on R1 to display the effect this change has made on the routing table. All OSPF routes for R1 are now being routed through R2.

```
R1# show ip route ospf
Codes: L - local, C - connected, S - static, R - RIP, M - mobile, B - BGP

        D - EIGRP, EX - EIGRP external, O - OSPF, IA - OSPF inter area

        N1 - OSPF NSSA external type 1, N2 - OSPF NSSA external type 2

        E1 - OSPF external type 1, E2 - OSPF external type 2

        i - IS-IS, su - IS-IS summary, L1 - IS-IS level-1, L2 - IS-IS level-2

        ia - IS-IS inter area, * - candidate default, U - per-user static route

        o - ODR, P - periodic downloaded static route, H - NHRP, l - LISP

        + - replicated route, % - next hop override


Gateway of last resort is not set


O       192.168.2.0/24 [110/782] via 192.168.12.2, 00:02:06, Serial0/0/0
O       192.168.3.0/24 [110/1563] via 192.168.12.2, 00:05:31, Serial0/0/0
        192.168.23.0/30 is subnetted, 1 subnets
O          192.168.23.0 [110/1562] via 192.168.12.2, 01:14:02, Serial0/0/0
```

Note: Manipulating link costs using the **ip ospf cost** command is the easiest and preferred method for changing OSPF route costs. In addition to changing the cost based on bandwidth, a network administrator may have other reasons for changing the cost of a route, such as preference for a particular service provider or the actual monetary cost of a link or route.

Explain why the route to the 192.168.3.0/24 network on R1 is now going through R2?

_____

_____

_____

Reflection

1. Why is it important to control the router ID assignment when using the OSPF protocol?

_____

_____

_____

_____

2. Why is the DR/BDR election process not a concern in this lab?

_____

_____

_____

3. Why would you want to set an OSPF interface to passive?

_____

_____

_____

Router Interface Summary Table

Router Interface Summary				
Router Model	**Ethernet Interface #1**	**Ethernet Interface #2**	**Serial Interface #1**	**Serial Interface #2**
1800	Fast Ethernet 0/0 (F0/0)	Fast Ethernet 0/1 (F0/1)	Serial 0/0/0 (S0/0/0)	Serial 0/0/1 (S0/0/1)
1900	Gigabit Ethernet 0/0 (G0/0)	Gigabit Ethernet 0/1 (G0/1)	Serial 0/0/0 (S0/0/0)	Serial 0/0/1 (S0/0/1)
2801	Fast Ethernet 0/0 (F0/0)	Fast Ethernet 0/1 (F0/1)	Serial 0/1/0 (S0/1/0)	Serial 0/1/1 (S0/1/1)
2811	Fast Ethernet 0/0 (F0/0)	Fast Ethernet 0/1 (F0/1)	Serial 0/0/0 (S0/0/0)	Serial 0/0/1 (S0/0/1)
2900	Gigabit Ethernet 0/0 (G0/0)	Gigabit Ethernet 0/1 (G0/1)	Serial 0/0/0 (S0/0/0)	Serial 0/0/1 (S0/0/1)

Note: To find out how the router is configured, look at the interfaces to identify the type of router and how many interfaces the router has. There is no way to effectively list all the combinations of configurations for each router class. This table includes identifiers for the possible combinations of Ethernet and Serial interfaces in the device. The table does not include any other type of interface, even though a specific router may contain one. An example of this might be an ISDN BRI interface. The string in parenthesis is the legal abbreviation that can be used in Cisco IOS commands to represent the interface.

8.3.3.6 Lab - Configuring Basic Single-Area OSPFv3

Topology

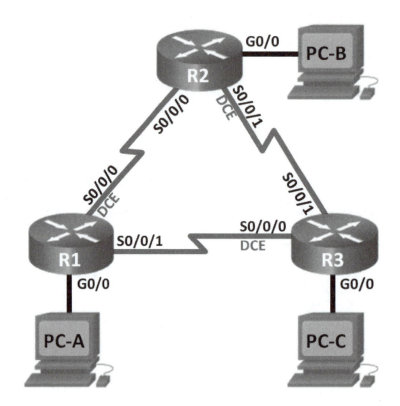

Addressing Table

Device	Interface	IPv6 Address	Default Gateway
R1	G0/0	2001:DB8:ACAD:A::1/64	
		FE80::1 link-local	N/A
	S0/0/0 (DCE)	2001:DB8:ACAD:12::1/64	
		FE80::1 link-local	N/A
	S0/0/1	2001:DB8:ACAD:13::1/64	
		FE80::1 link-local	N/A
R2	G0/0	2001:DB8:ACAD:B::2/64	
		FE80::2 link-local	N/A
	S0/0/0	2001:DB8:ACAD:12::2/64	
		FE80::2 link-local	N/A
	S0/0/1 (DCE)	2001:DB8:ACAD:23::2/64	
		FE80::2 link-local	N/A
R3	G0/0	2001:DB8:ACAD:C::3/64	
		FE80::3 link-local	N/A
	S0/0/0 (DCE)	2001:DB8:ACAD:13::3/64	
		FE80::3 link-local	N/A
	S0/0/1	2001:DB8:ACAD:23::3/64	
		FE80::3 link-local	N/A
PC-A	NIC	2001:DB8:ACAD:A::A/64	FE80::1
PC-B	NIC	2001:DB8:ACAD:B::B/64	FE80::2
PC-C	NIC	2001:DB8:ACAD:C::C/64	FE80::3

Objectives

Part 1: Build the Network and Configure Basic Device Settings

Part 2: Configure and Verify OSPFv3 Routing

Part 3: Configure OSPFv3 Passive Interfaces

Background / Scenario

Open Shortest Path First (OSPF) is a link-state routing protocol for IP networks. OSPFv2 is defined for IPv4 networks, and OSPFv3 is defined for IPv6 networks.

In this lab, you will configure the network topology with OSPFv3 routing, assign router IDs, configure passive interfaces, and use a number of CLI commands to display and verify OSPFv3 routing information.

Note: The routers used with CCNA hands-on labs are Cisco 1941 Integrated Services Routers (ISRs) with Cisco IOS Release 15.2(4)M3 (universalk9 image). Other routers and Cisco IOS versions can be used. Depending on the model and Cisco IOS version, the commands available and output produced might vary from what is shown in the labs. Refer to the Router Interface Summary Table at the end of this lab for the correct interface identifiers.

Note: Make sure that the routers have been erased and have no startup configurations. If you are unsure, contact your instructor.

Required Resources

- 3 Routers (Cisco 1941 with Cisco IOS Release 15.2(4)M3 universal image or comparable)
- 3 PCs (Windows 7, Vista, or XP with terminal emulation program, such as Tera Term)
- Console cables to configure the Cisco IOS devices via the console ports
- Ethernet and serial cables as shown in the topology

Part 1:   Build the Network and Configure Basic Device Settings

In Part 1, you will set up the network topology and configure basic settings on the PC hosts and routers.

Step 1:   Cable the network as shown in the topology.

Step 2:   Initialize and reload the routers as necessary.

Step 3:   Configure basic settings for each router.

a.   Disable DNS lookup.

b.   Configure device name as shown in the topology.

c.   Assign **class** as the privileged EXEC password.

d.   Assign **cisco** as the vty password.

e.   Configure a MOTD banner to warn users that unauthorized access is prohibited.

f.   Configure **logging synchronous** for the console line.

g.   Encrypt plain text passwords.

h.   Configure the IPv6 unicast and link-local addresses listed in the Addressing Table for all interfaces.

i.   Enable IPv6 unicast routing on each router.

j.   Copy the running configuration to the startup configuration.

Step 4:   Configure PC hosts.

Step 5:   Test connectivity.

The routers should be able to ping one another, and each PC should be able to ping its default gateway. The PCs are unable to ping other PCs until OSPFv3 routing is configured. Verify and troubleshoot if necessary.

Part 2:   Configure OSPFv3 Routing

In Part 2, you will configure OSPFv3 routing on all routers in the network and then verify that routing tables are updated correctly.

Step 1:   Assign router IDs.

OSPFv3 continues to use a 32 bit address for the router ID. Because there are no IPv4 addresses configured on the routers, you will manually assign the router ID using the **router-id** command.

a. Issue the **ipv6 router ospf** command to start an OSPFv3 process to the router.

   R1(config)# `ipv6 router ospf 1`

   Note: The OSPF process ID is kept locally and has no meaning to other routers on the network.

b. Assign the OSPFv3 router ID **1.1.1.1** to the R1.

   R1(config-rtr)# `router-id 1.1.1.1`

c. Start the OSPFv3 routing process and assign a router ID of **2.2.2.2** to R2 and a router ID of **3.3.3.3** to R3.

d. Issue the **show ipv6 ospf** command to verify the router IDs on all routers.

   R2# `show ipv6 ospf`
   ```
   Routing Process "ospfv3 1" with ID 2.2.2.2
   Event-log enabled, Maximum number of events: 1000, Mode: cyclic
   Router is not originating router-LSAs with maximum metric
   <output omitted>
   ```

Step 2: Configure OSPFv6 on R1.

With IPv6, it is common to have multiple IPv6 addresses configured on an interface. The network statement has been eliminated in OSPFv3. OSPFv3 routing is enabled at the interface level instead.

a. Issue the **ipv6 ospf 1 area 0** command for each interface on R1 that is to participate in OSPFv3 routing.

   R1(config)# `interface g0/0`

   R1(config-if)# `ipv6 ospf 1 area 0`

   R1(config-if)# `interface s0/0/0`

   R1(config-if)# `ipv6 ospf 1 area 0`

   R1(config-if)# `interface s0/0/1`

   R1(config-if)# `ipv6 ospf 1 area 0`

   Note: The process ID must match the process ID you used in Step1a.

b. Assign the interfaces on R2 and R3 to OSPFv3 area 0. You should see neighbor adjacency messages display when adding the interfaces to area 0.

   ```
   R1#
   *Mar 19 22:14:43.251: %OSPFv3-5-ADJCHG: Process 1, Nbr 2.2.2.2 on Serial0/0/0 from
   LOADING to FULL, Loading Done
   R1#
   *Mar 19 22:14:46.763: %OSPFv3-5-ADJCHG: Process 1, Nbr 3.3.3.3 on Serial0/0/1 from
   LOADING to FULL, Loading Done
   ```

Step 3: Verify OSPFv3 neighbors.

Issue the **show ipv6 ospf neighbor** command to verify that the router has formed an adjacency with its neighboring routers. If the router ID of the neighboring router is not displayed, or if its state does not show as FULL, the two routers have not formed an OSPF adjacency.

   R1# `show ipv6 ospf neighbor`

```
        OSPFv3 Router with ID (1.1.1.1) (Process ID 1)


   Neighbor ID     Pri   State           Dead Time   Interface ID   Interface
   3.3.3.3           0   FULL/  -        00:00:39    6              Serial0/0/1
   2.2.2.2           0   FULL/  -        00:00:36    6              Serial0/0/0
```

Step 4: **Verify OSPFv3 protocol settings.**

The **show ipv6 protocols** command is a quick way to verify vital OSPFv3 configuration information, including the OSPF process ID, the router ID, and the interfaces enabled for OSPFv3.

```
R1# show ipv6 protocols
IPv6 Routing Protocol is "connected"
IPv6 Routing Protocol is "ND"
IPv6 Routing Protocol is "ospf 1"
  Router ID 1.1.1.1
  Number of areas: 1 normal, 0 stub, 0 nssa
  Interfaces (Area 0):
    Serial0/0/1
    Serial0/0/0
    GigabitEthernet0/0
  Redistribution:
    None
```

Step 5: **Verify OSPFv3 interfaces.**

a. Issue the **show ipv6 ospf interface** command to display a detailed list for every OSPF-enabled interface.

```
R1# show ipv6 ospf interface
Serial0/0/1 is up, line protocol is up
  Link Local Address FE80::1, Interface ID 7
  Area 0, Process ID 1, Instance ID 0, Router ID 1.1.1.1
  Network Type POINT_TO_POINT, Cost: 64
  Transmit Delay is 1 sec, State POINT_TO_POINT
  Timer intervals configured, Hello 10, Dead 40, Wait 40, Retransmit 5
    Hello due in 00:00:05
  Graceful restart helper support enabled
  Index 1/3/3, flood queue length 0
  Next 0x0(0)/0x0(0)/0x0(0)
  Last flood scan length is 1, maximum is 1
  Last flood scan time is 0 msec, maximum is 0 msec
  Neighbor Count is 1, Adjacent neighbor count is 1
    Adjacent with neighbor 3.3.3.3
  Suppress hello for 0 neighbor(s)
Serial0/0/0 is up, line protocol is up
  Link Local Address FE80::1, Interface ID 6
  Area 0, Process ID 1, Instance ID 0, Router ID 1.1.1.1
  Network Type POINT_TO_POINT, Cost: 64
  Transmit Delay is 1 sec, State POINT_TO_POINT
  Timer intervals configured, Hello 10, Dead 40, Wait 40, Retransmit 5
    Hello due in 00:00:00
```

```
Graceful restart helper support enabled

Index 1/2/2, flood queue length 0

Next 0x0(0)/0x0(0)/0x0(0)

Last flood scan length is 1, maximum is 2

Last flood scan time is 0 msec, maximum is 0 msec

Neighbor Count is 1, Adjacent neighbor count is 1

  Adjacent with neighbor 2.2.2.2

Suppress hello for 0 neighbor(s)
GigabitEthernet0/0 is up, line protocol is up

Link Local Address FE80::1, Interface ID 3

Area 0, Process ID 1, Instance ID 0, Router ID 1.1.1.1

Network Type BROADCAST, Cost: 1

Transmit Delay is 1 sec, State DR, Priority 1

Designated Router (ID) 1.1.1.1, local address FE80::1

No backup designated router on this network

Timer intervals configured, Hello 10, Dead 40, Wait 40, Retransmit 5

  Hello due in 00:00:03

Graceful restart helper support enabled

Index 1/1/1, flood queue length 0

Next 0x0(0)/0x0(0)/0x0(0)

Last flood scan length is 0, maximum is 0

Last flood scan time is 0 msec, maximum is 0 msec

Neighbor Count is 0, Adjacent neighbor count is 0

Suppress hello for 0 neighbor(s)
```

b. To display a summary of OSPFv3-enabled interfaces, issue the **show ipv6 ospf interface brief** command.

```
R1# show ipv6 ospf interface brief
Interface    PID   Area          Intf ID   Cost   State  Nbrs F/C
Se0/0/1      1     0             7         64     P2P    1/1
Se0/0/0      1     0             6         64     P2P    1/1
Gi0/0        1     0             3         1      DR     0/0
```

Step 6: Verify the IPv6 routing table.

Issue the **show ipv6 route** command to verify that all networks are appearing in the routing table.

```
R2# show ipv6 route
IPv6 Routing Table - default - 10 entries
Codes: C - Connected, L - Local, S - Static, U - Per-user Static route
       B - BGP, R - RIP, I1 - ISIS L1, I2 - ISIS L2
       IA - ISIS interarea, IS - ISIS summary, D - EIGRP, EX - EIGRP external
       ND - ND Default, NDp - ND Prefix, DCE - Destination, NDr - Redirect
       O - OSPF Intra, OI - OSPF Inter, OE1 - OSPF ext 1, OE2 - OSPF ext 2
       ON1 - OSPF NSSA ext 1, ON2 - OSPF NSSA ext 2
```

```
O    2001:DB8:ACAD:A::/64 [110/65]
        via FE80::1, Serial0/0/0
C    2001:DB8:ACAD:B::/64 [0/0]
        via GigabitEthernet0/0, directly connected
L    2001:DB8:ACAD:B::2/128 [0/0]
        via GigabitEthernet0/0, receive
O    2001:DB8:ACAD:C::/64 [110/65]
        via FE80::3, Serial0/0/1
C    2001:DB8:ACAD:12::/64 [0/0]
        via Serial0/0/0, directly connected
L    2001:DB8:ACAD:12::2/128 [0/0]
        via Serial0/0/0, receive
O    2001:DB8:ACAD:13::/64 [110/128]
        via FE80::3, Serial0/0/1
        via FE80::1, Serial0/0/0
C    2001:DB8:ACAD:23::/64 [0/0]
        via Serial0/0/1, directly connected
L    2001:DB8:ACAD:23::2/128 [0/0]
        via Serial0/0/1, receive
L    FF00::/8 [0/0]
        via Null0, receive
```

What command would you use to only see the OSPF routes in the routing table?

Step 7:   **Verify end-to-end connectivity.**

Each PC should be able to ping the other PCs in the topology. Verify and troubleshoot if necessary.

Note: It may be necessary to disable the PC firewall to ping between PCs.

Part 3:   Configure OSPFv3 Passive Interfaces

The **passive-interface** command prevents routing updates from being sent through the specified router interface. This is commonly done to reduce traffic on the LANs as they do not need to receive dynamic routing protocol communication. In Part 3, you will use the **passive-interface** command to configure a single interface as passive. You will also configure OSPFv3 so that all interfaces on the router are passive by default, and then enable OSPF routing advertisements on selected interfaces.

Step 1:   Configure a passive interface.

a.   Issue the **show ipv6 ospf interface g0/0** command on R1. Notice the timer indicating when the next Hello packet is expected. Hello packets are sent every 10 seconds and are used between OSPF routers to verify that their neighbors are up.

```
R1# show ipv6 ospf interface g0/0
GigabitEthernet0/0 is up, line protocol is up
  Link Local Address FE80::1, Interface ID 3
  Area 0, Process ID 1, Instance ID 0, Router ID 1.1.1.1
  Network Type BROADCAST, Cost: 1
  Transmit Delay is 1 sec, State DR, Priority 1
  Designated Router (ID) 1.1.1.1, local address FE80::1
  No backup designated router on this network
  Timer intervals configured, Hello 10, Dead 40, Wait 40, Retransmit 5
    Hello due in 00:00:05
  Graceful restart helper support enabled
  Index 1/1/1, flood queue length 0
  Next 0x0(0)/0x0(0)/0x0(0)
  Last flood scan length is 0, maximum is 0
  Last flood scan time is 0 msec, maximum is 0 msec
  Neighbor Count is 0, Adjacent neighbor count is 0
  Suppress hello for 0 neighbor(s)
```

b. Issue the **passive-interface** command to change the G0/0 interface on R1 to passive.

```
R1(config)# ipv6 router ospf 1
R1(config-rtr)# passive-interface g0/0
```

c. Re-issue the **show ipv6 ospf interface g0/0** command to verify that G0/0 is now passive.

```
R1# show ipv6 ospf interface g0/0
GigabitEthernet0/0 is up, line protocol is up
  Link Local Address FE80::1, Interface ID 3
  Area 0, Process ID 1, Instance ID 0, Router ID 1.1.1.1
  Network Type BROADCAST, Cost: 1
  Transmit Delay is 1 sec, State WAITING, Priority 1
  No designated router on this network
  No backup designated router on this network
  Timer intervals configured, Hello 10, Dead 40, Wait 40, Retransmit 5
    No Hellos (Passive interface)
    Wait time before Designated router selection 00:00:34
  Graceful restart helper support enabled
  Index 1/1/1, flood queue length 0
  Next 0x0(0)/0x0(0)/0x0(0)
  Last flood scan length is 0, maximum is 0
  Last flood scan time is 0 msec, maximum is 0 msec
  Neighbor Count is 0, Adjacent neighbor count is 0
  Suppress hello for 0 neighbor(s)
```

d. Issue the **show ipv6 route ospf** command on R2 and R3 to verify that a route to the 2001:DB8:ACAD:A::/64 network is still available.

```
R2# show ipv6 route ospf
IPv6 Routing Table - default - 10 entries
Codes: C - Connected, L - Local, S - Static, U - Per-user Static route
       B - BGP, R - RIP, I1 - ISIS L1, I2 - ISIS L2
       IA - ISIS interarea, IS - ISIS summary, D - EIGRP, EX - EIGRP external
       ND - ND Default, NDp - ND Prefix, DCE - Destination, NDr - Redirect
       O - OSPF Intra, OI - OSPF Inter, OE1 - OSPF ext 1, OE2 - OSPF ext 2
       ON1 - OSPF NSSA ext 1, ON2 - OSPF NSSA ext 2
O   2001:DB8:ACAD:A::/64 [110/65]
     via FE80::1, Serial0/0/0
O   2001:DB8:ACAD:C::/64 [110/65]
     via FE80::3, Serial0/0/1
O   2001:DB8:ACAD:13::/64 [110/128]
     via FE80::3, Serial0/0/1
     via FE80::1, Serial0/0/0
```

Step 2: Set passive interface as the default on the router.

a. Issue the **passive-interface default** command on R2 to set the default for all OSPFv3 interfaces as passive.

```
R2(config)# ipv6 router ospf 1
R2(config-rtr)# passive-interface default
```

b. Issue the **show ipv6 ospf neighbor** command on R1. After the dead timer expires, R2 is no longer listed as an OSPF neighbor.

```
R1# show ipv6 ospf neighbor


       OSPFv3 Router with ID (1.1.1.1) (Process ID 1)


Neighbor ID     Pri   State          Dead Time   Interface ID    Interface
3.3.3.3           0   FULL/  -       00:00:37    6               Serial0/0/1
```

c. On R2, issue the **show ipv6 ospf interface s0/0/0** command to view the OSPF status of interface S0/0/0.

```
R2# show ipv6 ospf interface s0/0/0
Serial0/0/0 is up, line protocol is up
  Link Local Address FE80::2, Interface ID 6
  Area 0, Process ID 1, Instance ID 0, Router ID 2.2.2.2
  Network Type POINT_TO_POINT, Cost: 64
  Transmit Delay is 1 sec, State POINT_TO_POINT
  Timer intervals configured, Hello 10, Dead 40, Wait 40, Retransmit 5
    No Hellos (Passive interface)
  Graceful restart helper support enabled
```

```
Index 1/2/2, flood queue length 0

Next 0x0(0)/0x0(0)/0x0(0)

Last flood scan length is 2, maximum is 3

Last flood scan time is 0 msec, maximum is 0 msec

Neighbor Count is 0, Adjacent neighbor count is 0

Suppress hello for 0 neighbor(s)
```

d. If all OSPFv3 interfaces on R2 are passive, then no routing information is being advertised. If this is the case, then R1 and R3 should no longer have a route to the 2001:DB8:ACAD:B::/64 network. You can verify this by using the **show ipv6 route** command.

e. Change S0/0/1 on R2 by issuing the **no passive-interface** command, so that it sends and receives OSPFv3 routing updates. After entering this command, an informational message displays stating that a neighbor adjacency has been established with R3.

```
R2(config)# ipv6 router ospf 1

R2(config-rtr)# no passive-interface s0/0/1

*Apr  8 19:21:57.939: %OSPFv3-5-ADJCHG: Process 1, Nbr 3.3.3.3 on Serial0/0/1
from LOADING to FULL, Loading Done
```

f. Re-issue the **show ipv6 route** and **show ipv6 ospf neighbor** commands on R1 and R3, and look for a route to the 2001:DB8:ACAD:B::/64 network.

What interface is R1 using to route to the 2001:DB8:ACAD:B::/64 network? _____

What is the accumulated cost metric for the 2001:DB8:ACAD:B::/64 network on R1? _____

Does R2 show up as an OSPFv3 neighbor on R1? _____

Does R2 show up as an OSPFv3 neighbor on R3? _____

What does this information tell you?

_____

_____

_____

_____

g. On R2, issue the **no passive-interface S0/0/0** command to allow OSPFv3 routing updates to be advertised on that interface.

h. Verify that R1 and R2 are now OSPFv3 neighbors.

Reflection

1. If the OSPFv6 configuration for R1 had a process ID of 1, and the OSPFv3 configuration for R2 had a process ID of 2, can routing information be exchanged between the two routers? Why?

_____

_____

2. What may have been the reasoning for removing the **network** command in OSPFv3?

_____

_____

_____

Router Interface Summary Table

Router Interface Summary				
Router Model	**Ethernet Interface #1**	**Ethernet Interface #2**	**Serial Interface #1**	**Serial Interface #2**
1800	Fast Ethernet 0/0 (F0/0)	Fast Ethernet 0/1 (F0/1)	Serial 0/0/0 (S0/0/0)	Serial 0/0/1 (S0/0/1)
1900	Gigabit Ethernet 0/0 (G0/0)	Gigabit Ethernet 0/1 (G0/1)	Serial 0/0/0 (S0/0/0)	Serial 0/0/1 (S0/0/1)
2801	Fast Ethernet 0/0 (F0/0)	Fast Ethernet 0/1 (F0/1)	Serial 0/1/0 (S0/1/0)	Serial 0/1/1 (S0/1/1)
2811	Fast Ethernet 0/0 (F0/0)	Fast Ethernet 0/1 (F0/1)	Serial 0/0/0 (S0/0/0)	Serial 0/0/1 (S0/0/1)
2900	Gigabit Ethernet 0/0 (G0/0)	Gigabit Ethernet 0/1 (G0/1)	Serial 0/0/0 (S0/0/0)	Serial 0/0/1 (S0/0/1)
Note: To find out how the router is configured, look at the interfaces to identify the type of router and how many interfaces the router has. There is no way to effectively list all the combinations of configurations for each router class. This table includes identifiers for the possible combinations of Ethernet and Serial interfaces in the device. The table does not include any other type of interface, even though a specific router may contain one. An example of this might be an ISDN BRI interface. The string in parenthesis is the legal abbreviation that can be used in Cisco IOS commands to represent the interface.				

8.4.1.1 Class Activity – Stepping Through OSPFv3

Objectives

Explain the process by which link-state routers learn about other networks.

Scenario

This class activity is designed for groups of three students. The objective is to review the Shortest Path First (SPF) routing process.

You will design and address a network, communicate the network address scheme and operation of network links to your group members, and compute the SPF.

Complete the steps as shown on the PDF for this class activity. If you have time, share your network design and Open Shortest Path First (OSPF) process with another group.

Resources

In preparation of this activity, you will need two different IPv6 network and cost numbers. The IPv6 network numbers must be chosen with the following format: 2002:DB8:AAAA:?::0/64, where **?** is a student-selected network number. You have two choices for *cost* – 10 (Fast Ethernet network), or 1 (Gigabit Ethernet network).

Bring your two IPv6 network and cost numbers to the group setting. One student in your group will act as the recorder, will draw three circles, and connect them on paper. Each circle will represent a student's router and the connecting lines will represent the networks and links to be agreed upon.

Each group member should follow Steps 1 to 4 (below) in the order listed. As the group progresses through the activity, you should keep personal notes about your own router, including information about neighbor adjacency, link-state advertisements, topology table entries, and the SPF algorithm.

Directions

Step 1:

a. Speak to the classmate to your left. Compare network and cost numbers brought to the group. Agree upon an IPv6 network, links, and cost numbers you would like to use between your two routers. Remember, you may only use 1 (Gigabit Ethernet) or 10 (Fast Ethernet) for cost. When you have agreed upon your network, link numbers, and determined the cost of the route, record the information on the paper graphic created by the recorder.

b. Complete the same process with the classmate to your right.

c. After speaking with both of your direct neighbors, you have agreed upon two networks with link addresses and the cost of the route. Record the information you agreed upon on the paper graphic.

Step 2:

    a.   Each student will speak only to their direct neighbors. They will share all of their IPv6 network and link numbers and the cost of the networks to which they are connected. Almost immediately, everyone in the group will know about all networks, their links, and the cost of the individual networks between neighbors.

    b.   Check with the group members to ascertain all group members have the same information with which to work for Step 3.

Step 3:

    a.   On your own paper, create a table listing possible paths to all other networks. Use the formula supplied with this chapter $n(n - 1)/2$. You will have a total of four possible routes to list on your table.

    b.   On the table created in the Step 3 a., add a column with the headings, IPv6 Network Number and Cost.

    c.   Fill in the table with information you know about the networks on your group's topology.

Step 4:

    a.   Go back to the table created in Step 3.

    b.   Place a star by the lowest-cost routes to all other routers.

When these four steps are complete, you have established neighbor adjacencies, exchanged link-state advertisements, built a topology table, and created a routing table with the best cost to all other networks within your group or area.

If you have the time, refer to your topology table and build the network on real equipment or Packet Tracer. Use some or all of the commands listed below to prove OSPF's operation:

```
R1# show ipv6 interface brief
R1# show ipv6 protocols
R1# show ip protocols
R1# show ipv6 route
```

Reflection

1.   Which OSPFv3 processing step is reviewed in Step 1 of this activity?

_____

2.   Which OSPFv3 processing step is reviewed in Step 2 of this activity?

_____

3.   Which process for OSPFv3 is reviewed in Step 3 of this activity?

_____

4.   Which process step for OSPFv3 is reviewed in Step 4 of this activity?

_____

Chapter 9 — Access Control Lists

9.0.1.2 Class Activity – Permit Me to Assist You

Objective

Explain the purpose and operation of ACLs.

Scenario

- Each individual in the class will record five questions they would ask a candidate who is applying for a security clearance for a network assistant position within a small- to medium-sized business. The list of questions should be listed in order of importance to selecting a good candidate for the job. The preferred answers will also be recorded.

- After three minutes of brainstorming the list of questions, the instructor will ask two students to serve as interviewers. These two students will use only their list of questions and answers for the next part of this activity. The instructor will explain to only the two interviewers that they have the discretion, at any time, to stop the process and state "you are all permitted to the next level of interviews" or "I am sorry, but you do not have the qualifications to continue to the next level of interviews." The interviewer does not need to complete all of the questions on the list.

- The rest of the class will be split in half and assigned to one of the interviewers.

- Once everyone is settled into their group with an interviewer, the group application interviews will begin.

- The two selected interviewers will ask the first question on the list that they created; an example would be "are you over the age of 18?" If the applicant does not meet the age requirement, as specified by the interviewer's original questions and answers, the applicant will be eliminated from the pool of applicants and must move to another area within the room where they will observe the rest of the application process.

The next question will then be asked by the interviewer. If applicants answer correctly, they may stay with the applicant group. The entire class will then get together and discuss their observations regarding the process to permit or deny them the opportunity to continue on to the next level of interviews.

Reflection

1. What factors did you consider when devising your list of criteria for network assistant security clearance?

_____

2. How difficult was it to devise five security questions to deliver during the interviews? Why were you asked to list your questions in order of importance to selecting a good candidate?

_____

3. Why would the process of elimination be stopped, even if there were still a few applicants available?

_____

4. How could this scenario and the results be applied to network traffic?

_____

9.2.2.7 Lab – Configuring and Verifying Standard ACLs

Topology

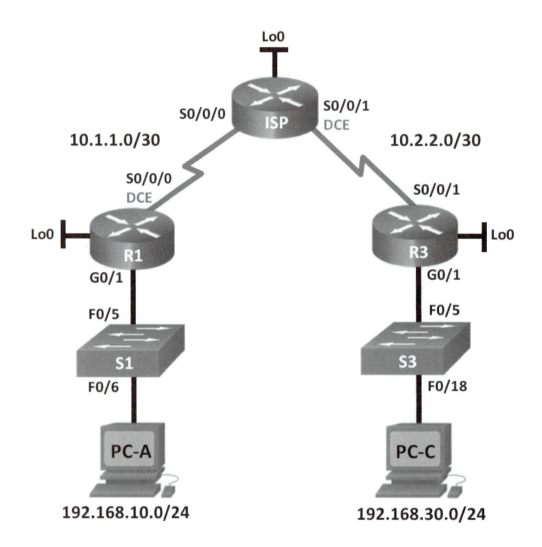

Addressing Table

Device	Interface	IP Address	Subnet Mask	Default Gateway
R1	G0/1	192.168.10.1	255.255.255.0	N/A
	Lo0	192.168.20.1	255.255.255.0	N/A
	S0/0/0 (DCE)	10.1.1.1	255.255.255.252	N/A
ISP	S0/0/0	10.1.1.2	255.255.255.252	N/A
	S0/0/1 (DCE)	10.2.2.2	255.255.255.252	N/A
	Lo0	209.165.200.225	255.255.255.224	N/A
R3	G0/1	192.168.30.1	255.255.255.0	N/A
	Lo0	192.168.40.1	255.255.255.0	N/A
	S0/0/1	10.2.2.1	255.255.255.252	N/A
S1	VLAN 1	192.168.10.11	255.255.255.0	192.168.10.1
S3	VLAN 1	192.168.30.11	255.255.255.0	192.168.30.1
PC-A	NIC	192.168.10.3	255.255.255.0	192.168.10.1
PC-C	NIC	192.168.30.3	255.255.255.0	192.168.30.1

Objectives

Part 1: Set Up the Topology and Initialize Devices

- Set up equipment to match the network topology.
- Initialize and reload the routers and switches.

Part 2: Configure Devices and Verify Connectivity

- Assign a static IP address to PCs.
- Configure basic settings on routers.
- Configure basic settings on switches.
- Configure EIGRP routing on R1, ISP, and R3.
- Verify connectivity between devices.

Part 3: Configure and Verify Standard Numbered and Named ACLs

- Configure, apply, and verify a numbered standard ACL.
- Configure, apply, and verify a named ACL.

Part 4: Modify a Standard ACL

- Modify and verify a named standard ACL.
- Test the ACL.

Background / Scenario

Network security is an important issue when designing and managing IP networks. The ability to configure proper rules to filter packets, based on established security policies, is a valuable skill.

In this lab, you will set up filtering rules for two offices represented by R1 and R3. Management has established some access policies between the LANs located at R1 and R3, which you must implement. The ISP router sitting between R1 and R3 will not have any ACLs placed on it. You would not be allowed any administrative access to an ISP router because you can only control and manage your own equipment.

Note: The routers used with CCNA hands-on labs are Cisco 1941 Integrated Services Routers (ISRs) with Cisco IOS Release 15.2(4)M3 (universalk9 image). The switches used are Cisco Catalyst 2960s with Cisco IOS Release 15.0(2) (lanbasek9 image). Other routers, switches, and Cisco IOS versions can be used. Depending on the model and Cisco IOS version, the commands available and output produced might vary from what is shown in the labs. Refer to the Router Interface Summary Table at the end of the lab for the correct interface identifiers.

Note: Make sure that the routers and switches have been erased and have no startup configurations. If you are unsure, contact your instructor.

Required Resources

- 3 Routers (Cisco 1941 with Cisco IOS Release 15.2(4)M3 universal image or comparable)
- 2 Switches (Cisco 2960 with Cisco IOS Release 15.0(2) lanbasek9 image or comparable)
- 2 PCs (Windows 7, Vista, or XP with terminal emulation program, such as Tera Term)
- Console cables to configure the Cisco IOS devices via the console ports
- Ethernet and serial cables as shown in the topology

Part 1: Set Up the Topology and Initialize Devices

In Part 1, you set up the network topology and clear any configurations, if necessary.

Step 1: Cable the network as shown in the topology.

Step 2: Initialize and reload the routers and switches.

Part 2: Configure Devices and Verify Connectivity

In Part 2, you configure basic settings on the routers, switches, and PCs. Refer to the Topology and Addressing Table for device names and address information.

Step 1: Configure IP addresses on PC-A and PC-C.

Step 2: Configure basic settings for the routers.

a. Disable DNS lookup.

b. Configure the device names as shown in the topology.

c. Create loopback interfaces on each router as shown in the Addressing Table.

d. Configure interface IP addresses as shown in the Topology and Addressing Table.

e. Configure a privileged EXEC mode password of **class**.

f. Assign a clock rate of **128000** to the DCE serial interfaces.

g. Assign **cisco** as the console password.

h. Assign **cisco** as the vty password and enable Telnet access.

Step 3:   (Optional) Configure basic settings on the switches.

a.   Disable DNS lookup.

b.   Configure the device names as shown in the topology.

c.   Configure the management interface IP address as shown in the Topology and Addressing Table.

d.   Configure a privileged EXEC mode password of **class**.

e.   Configure a default gateway.

f.   Assign **cisco** as the console password.

g.   Assign **cisco** as the vty password and enable Telnet access.

Step 4:   Configure EIGRP routing on R1, ISP, and R3.

a.   Configure autonomous system (AS) number 10 and advertise all networks on R1, ISP, and R3. Disable automatic summarization.

b.   After configuring EIGRP on R1, ISP, and R3, verify that all routers have complete routing tables listing all networks. Troubleshoot if this is not the case.

Step 5:   Verify connectivity between devices.

Note: It is very important to test whether connectivity is working **before** you configure and apply access lists! You want to ensure that your network is properly functioning before you start to filter traffic.

a.   From PC-A, ping PC-C and the loopback interface on R3. Were your pings successful? _____

b.   From R1, ping PC-C and the loopback interface on R3. Were your pings successful? _____

c.   From PC-C, ping PC-A and the loopback interface on R1. Were your pings successful? _____

d.   From R3, ping PC-A and the loopback interface on R1. Were your pings successful? _____

Part 3:   Configure and Verify Standard Numbered and Named ACLs

Step 1:   Configure a numbered standard ACL.

Standard ACLs filter traffic based on the source IP address only. A typical best practice for standard ACLs is to configure and apply it as close to the destination as possible. For the first access list, create a standard numbered ACL that allows traffic from all hosts on the 192.168.10.0/24 network and all hosts on the 192.168.20.0/24 network to access all hosts on the 192.168.30.0/24 network. The security policy also states that a **deny any** access control entry (ACE), also referred to as an ACL statement, should be present at the end of all ACLs.

What wildcard mask would you use to allow all hosts on the 192.168.10.0/24 network to access the 192.168.30.0/24 network?

_____

Following Cisco's recommended best practices, on which router would you place this ACL? _____

On which interface would you place this ACL? In what direction would you apply it?

_____

a. Configure the ACL on R3. Use 1 for the access list number.

    R3(config)# **access-list 1 remark Allow R1 LANs Access**

    R3(config)# **access-list 1 permit 192.168.10.0 0.0.0.255**

    R3(config)# **access-list 1 permit 192.168.20.0 0.0.0.255**

    R3(config)# **access-list 1 deny any**

b. Apply the ACL to the appropriate interface in the proper direction.

    R3(config)# **interface g0/1**

    R3(config-if)# **ip access-group 1 out**

c. Verify a numbered ACL.

    The use of various **show** commands can aid you in verifying both the syntax and placement of your ACLs in your router.

    To see access list 1 in its entirety with all ACEs, which command would you use?

    _____

    What command would you use to see where the access list was applied and in what direction?

    _____

    1) On R3, issue the **show access-lists 1** command.

        R3# **show access-list 1**

        Standard IP access list 1

            10 permit 192.168.10.0, wildcard bits 0.0.0.255

            20 permit 192.168.20.0, wildcard bits 0.0.0.255

            30 deny   any

    2) On R3, issue the **show ip interface g0/1** command.

        R3# **show ip interface g0/1**

        GigabitEthernet0/1 is up, line protocol is up

          Internet address is 192.168.30.1/24

          Broadcast address is 255.255.255.255

          Address determined by non-volatile memory

          MTU is 1500 bytes

          Helper address is not set

          Directed broadcast forwarding is disabled

          Multicast reserved groups joined: 224.0.0.10

          Outgoing access list is 1

          Inbound access list is not set

          Output omitted

    3) Test the ACL to see if it allows traffic from the 192.168.10.0/24 network access to the 192.168.30.0/24 network. From the PC-A command prompt, ping the PC-C IP address. Were the pings successful? _____

    4) Test the ACL to see if it allows traffic from the 192.168.20.0/24 network access to the 192.168.30.0/24 network. You must do an extended ping and use the loopback 0 address on R1 as your source. Ping PC-C's IP address. Were the pings successful? _____

```
R1# ping

Protocol [ip]:

Target IP address: 192.168.30.3

Repeat count [5]:

Datagram size [100]:

Timeout in seconds [2]:

Extended commands [n]: y

Source address or interface: 192.168.20.1

Type of service [0]:

Set DF bit in IP header? [no]:

Validate reply data? [no]:

Data pattern [0xABCD]:

Loose, Strict, Record, Timestamp, Verbose[none]:

Sweep range of sizes [n]:

Type escape sequence to abort.

Sending 5, 100-byte ICMP Echos to 192.168.30.3, timeout is 2 seconds:

Packet sent with a source address of 192.168.20.1

!!!!!

Success rate is 100 percent (5/5), round-trip min/avg/max = 28/29/32 ms
```

d. From the R1 prompt, ping PC-C's IP address again.

```
R1# ping 192.168.3.3
```

Was the ping successful? Why or why not?

_____

_____

_____

Step 2:   **Configure a named standard ACL.**

Create a named standard ACL that conforms to the following policy: allow traffic from all hosts on the 192.168.40.0/24 network access to all hosts on the 192.168.10.0/24 network. Also, only allow host PC-C access to the 192.168.10.0/24 network. The name of this access list should be called BRANCH-OFFICE-POLICY.

Following Cisco's recommended best practices, on which router would you place this ACL? _____

On which interface would you place this ACL? In what direction would you apply it?

_____

a. Create the standard named ACL BRANCH-OFFICE-POLICY on R1.

    ```
    R1(config)# ip access-list standard BRANCH-OFFICE-POLICY
    R1(config-std-nacl)# permit host 192.168.30.3
    R1(config-std-nacl)# permit 192.168.40.0 0.0.0.255
    R1(config-std-nacl)# end
    R1#
    *Feb 15 15:56:55.707: %SYS-5-CONFIG_I: Configured from console by console
    ```

    Looking at the first permit ACE in the access list, what is another way to write this?

    _____

b. Apply the ACL to the appropriate interface in the proper direction.

    ```
    R1# config t
    R1(config)# interface g0/1
    R1(config-if)# ip access-group BRANCH-OFFICE-POLICY out
    ```

c. Verify a named ACL.

    1) On R1, issue the **show access-lists** command.

        ```
        R1# show access-lists
        Standard IP access list BRANCH-OFFICE-POLICY
            10 permit 192.168.30.3
            20 permit 192.168.40.0, wildcard bits 0.0.0.255
        ```

        Is there any difference between this ACL on R1 with the ACL on R3? If so, what is it?

        _____

        _____

        _____

        _____

        _____

    2) On R1, issue the **show ip interface g0/1** command.
        ```
        R1# show ip interface g0/1
        GigabitEthernet0/1 is up, line protocol is up
          Internet address is 192.168.10.1/24
          Broadcast address is 255.255.255.255
          Address determined by non-volatile memory
          MTU is 1500 bytes
          Helper address is not set
          Directed broadcast forwarding is disabled
          Multicast reserved groups joined: 224.0.0.10
          Outgoing access list is BRANCH-OFFICE-POLICY
        ```

```
    Inbound access list is not set
<Output omitted>
```

3) Test the ACL. From the command prompt on PC-C, ping PC-A's IP address. Were the pings successful? _____

4) Test the ACL to ensure that only the PC-C host is allowed access to the 192.168.10.0/24 network. You must do an extended ping and use the G0/1 address on R3 as your source. Ping PC-A's IP address. Were the pings successful? _____

```
R3# ping

Protocol [ip]:

Target IP address: 192.168.10.3

Repeat count [5]:

Datagram size [100]:

Timeout in seconds [2]:

Extended commands [n]: y

Source address or interface: 192.168.30.1

Type of service [0]:

Set DF bit in IP header? [no]:

Validate reply data? [no]:

Data pattern [0xABCD]:

Loose, Strict, Record, Timestamp, Verbose[none]:

Sweep range of sizes [n]:

Type escape sequence to abort.

Sending 5, 100-byte ICMP Echos to 192.168.10.3, timeout is 2 seconds:

Packet sent with a source address of 192.168.30.1

U.U.U
```

5) Test the ACL to see if it allows traffic from the 192.168.40.0/24 network access to the 192.168.10.0/24 network. You must perform an extended ping and use the loopback 0 address on R3 as your source. Ping PC-A's IP address. Were the pings successful? _____

Part 4:   Modify a Standard ACL

It is common in business for security policies to change. For this reason, ACLs may need to be modified. In Part 4, you will change one of the previous ACLs you configured, to match a new management policy being put in place.

Management has decided that users from the 209.165.200.224/27 network should be allowed full access to the 192.168.10.0/24 network. Management also wants ACLs on all of their routers to follow consistent rules. A **deny any** ACE should be placed at the end of all ACLs. You must modify the BRANCH-OFFICE-POLICY ACL.

You will add two additional lines to this ACL. There are two ways you could do this:

OPTION 1: Issue a **no ip access-list standard BRANCH-OFFICE-POLICY** command in global configuration mode. This would effectively take the whole ACL out of the router. Depending upon the router IOS, one of the following scenarios would occur: all filtering of packets would be cancelled and all packets would be allowed through the router; or, because you did not take off the **ip access-group** command on the G0/1 interface, filtering is still in place. Regardless, when the ACL is gone, you could retype the whole ACL, or cut and paste it in from a text editor.

OPTION 2: You can modify ACLs in place by adding or deleting specific lines within the ACL itself. This can come in handy, especially with ACLs that have many lines of code. The retyping of the whole ACL or cutting and pasting can easily lead to errors. Modifying specific lines within the ACL is easily accomplished.

Note: For this lab, use Option 2.

Step 1:   Modify a named standard ACL.

a.   From R1 privilege EXEC mode, issue a **show access-lists** command.

```
R1# show access-lists
Standard IP access list BRANCH-OFFICE-POLICY
    10 permit 192.168.30.3 (8 matches)
    20 permit 192.168.40.0, wildcard bits 0.0.0.255 (5 matches)
```

b.   Add two additional lines at the end of the ACL. From global config mode, modify the ACL, BRANCH-OFFICE-POLICY.

```
R1#(config)# ip access-list standard BRANCH-OFFICE-POLICY
R1(config-std-nacl)# 30 permit 209.165.200.224 0.0.0.31
R1(config-std-nacl)# 40 deny any
R1(config-std-nacl)# end
```

c.   Verify the ACL.

1)   On R1, issue the **show access-lists** command.

```
R1# show access-lists
Standard IP access list BRANCH-OFFICE-POLICY
    10 permit 192.168.30.3 (8 matches)
    20 permit 192.168.40.0, wildcard bits 0.0.0.255 (5 matches)
    30 permit 209.165.200.224, wildcard bits 0.0.0.31
    40 deny    any
```

Do you have to apply the BRANCH-OFFICE-POLICY to the G0/1 interface on R1?

_____

_____

2) From the ISP command prompt, issue an extended ping. Test the ACL to see if it allows traffic from the 209.165.200.224/27 network access to the 192.168.10.0/24 network. You must do an extended ping and use the loopback 0 address on ISP as your source. Ping PC-A's IP address. Were the pings successful? _____

Reflection

1. As you can see, standard ACLs are very powerful and work quite well. Why would you ever have the need for using extended ACLs?

_____

_____

_____

_____

2. Typically, more typing is required when using a named ACL as opposed to a numbered ACL. Why would you choose named ACLs over numbered?

_____

_____

_____

_____

Router Interface Summary Table

Router Interface Summary				
Router Model	**Ethernet Interface #1**	**Ethernet Interface #2**	**Serial Interface #1**	**Serial Interface #2**
1800	Fast Ethernet 0/0 (F0/0)	Fast Ethernet 0/1 (F0/1)	Serial 0/0/0 (S0/0/0)	Serial 0/0/1 (S0/0/1)
1900	Gigabit Ethernet 0/0 (G0/0)	Gigabit Ethernet 0/1 (G0/1)	Serial 0/0/0 (S0/0/0)	Serial 0/0/1 (S0/0/1)
2801	Fast Ethernet 0/0 (F0/0)	Fast Ethernet 0/1 (F0/1)	Serial 0/1/0 (S0/1/0)	Serial 0/1/1 (S0/1/1)
2811	Fast Ethernet 0/0 (F0/0)	Fast Ethernet 0/1 (F0/1)	Serial 0/0/0 (S0/0/0)	Serial 0/0/1 (S0/0/1)
2900	Gigabit Ethernet 0/0 (G0/0)	Gigabit Ethernet 0/1 (G0/1)	Serial 0/0/0 (S0/0/0)	Serial 0/0/1 (S0/0/1)
Note: To find out how the router is configured, look at the interfaces to identify the type of router and how many interfaces the router has. There is no way to effectively list all the combinations of configurations for each router class. This table includes identifiers for the possible combinations of Ethernet and Serial interfaces in the device. The table does not include any other type of interface, even though a specific router may contain one. An example of this might be an ISDN BRI interface. The string in parenthesis is the legal abbreviation that can be used in Cisco IOS commands to represent the interface.				

9.2.3.4 Lab – Configuring and Verifying VTY Restrictions

Topology

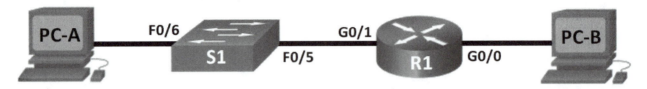

Addressing Table

Device	Interface	IP Address	Subnet Mask	Default Gateway
R1	G0/0	192.168.0.1	255.255.255.0	N/A
	G0/1	192.168.1.1	255.255.255.0	N/A
S1	VLAN 1	192.168.1.2	255.255.255.0	192.168.1.1
PC-A	NIC	192.168.1.3	255.255.255.0	192.168.1.1
PC-B	NIC	192.168.0.3	255.255.255.0	192.168.0.1

Objectives

Part 1: Configure Basic Device Settings

Part 2: Configure and Apply the Access Control List on R1

Part 3: Verify the Access Control List Using Telnet

Part 4: Challenge - Configure and Apply the Access Control List on S1

Background / Scenario

It is a good practice to restrict access to the router management interfaces, such as the console and vty lines. An access control list (ACL) can be used to allow access for specific IP addresses, ensuring that only the administrator PCs have permission to telnet or SSH into the router.

Note: In the Cisco device outputs, ACL are abbreviated as access-list.

In this lab, you will create and apply a named standard ACL to restrict remote access to the router vty lines.

After the ACL has been created and applied, you will test and verify the ACL by accessing the router from different IP addresses using Telnet.

This lab will provide the commands necessary for creating and applying the ACL.

Note: The routers used with CCNA hands-on labs are Cisco 1941 Integrated Services Routers (ISRs) with Cisco IOS Release 15.2(4)M3 (universalk9 image). The switches used are Cisco Catalyst 2960s with Cisco IOS Release 15.0(2) (lanbasek9 image). Other routers, switches, and Cisco IOS versions can be used. Depending on the model and Cisco IOS version, the commands available and output produced might vary from what is shown in the labs. Refer to the Router Interface Summary Table at the end of the lab for the correct interface identifiers.

Note: Make sure that the routers and switches have been erased and have no startup configurations. If you are unsure, contact your instructor.

Required Resources

- 1 Router (Cisco 1941 with Cisco IOS Release 15.2(4)M3 universal image or comparable)
- 1 Switch (Cisco 2960 with Cisco IOS Release 15.0(2) lanbasek9 image or comparable)
- 2 PCs (Windows 7, Vista, or XP with terminal emulation program, such as Tera Term)
- Console cables to configure the Cisco IOS devices via the console ports
- Ethernet cables as shown in the topology

Note: The Gigabit Ethernet interfaces on Cisco 1941 routers are autosensing and an Ethernet straight-through cable may be used between the router and PC-B. If using another model Cisco router, it may be necessary to use an Ethernet crossover cable.

Part 1: Configure Basic Device Settings

In Part 1, you will set up the network topology and configure the interface IP addresses, device access, and passwords on the router.

Step 1:   Cable the network as shown in the topology diagram.

Step 2:   Configure the PC-A and PC-B network settings according to the Addressing Table.

Step 3:   Initialize and reload the router and switch.

a. Disable DNS lookup.

b. Configure device names according to the Topology diagram.

c. Assign **class** as the privileged EXEC encrypted password.

d. Assign **cisco** as the console password, activate logging synchronous, and enable login.

e. Assign **cisco** as the vty password, activate logging synchronous, and enable login.

f. Encrypt the plain text passwords.

g. Create a banner that warns anyone accessing the device that unauthorized access is prohibited.

h. Configure IP addresses on the interfaces listed in the Addressing Table.

i. Configure the default gateway for the switch.

j. Save the running configuration to the startup configuration file.

Part 2: Configure and Apply the Access Control List on R1

In Part 2, you will configure a named standard ACL and apply it to the router virtual terminal lines to restrict remote access to the router.

Step 1:   Configure and apply a standard named ACL.

a. Console into the router R1 and enable privileged EXEC mode.

b. From global configuration mode, view the command options under **ip access-list** by using a space and a question mark.

```
R1(config)# ip access-list ?

   extended     Extended Access List

   helper       Access List acts on helper-address

   log-update   Control access list log updates
```

```
logging     Control access list logging

resequence  Resequence Access List

standard    Standard Access List
```

c. View the command options under **ip access-list standard** by using a space and a question mark.

```
R1(config)# ip access-list standard ?

<1-99>       Standard IP access-list number

<1300-1999>  Standard IP access-list number (expanded range)

WORD         Access-list name
```

d. Add **ADMIN-MGT** to the end of the **ip access-list standard** command and press Enter. You are now in the standard named access-list configuration mode (config-std-nacl).

```
R1(config)# ip access-list standard ADMIN-MGT

R1(config-std-nacl)#
```

e. Enter your ACL permit or deny access control entry (ACE), also known as an ACL statement, one line at a time. Remember that there is an implicit **deny any** at the end of the ACL, which effectively denies all traffic. Enter a question mark to view your command options.

```
R1(config-std-nacl)# ?

Standard Access List configuration commands:

  <1-2147483647>  Sequence Number

  default         Set a command to its defaults

  deny            Specify packets to reject

  exit            Exit from access-list configuration mode

  no              Negate a command or set its defaults

  permit          Specify packets to forward

  remark          Access list entry comment
```

f. Create a permit ACE for Administrator PC-A at 192.168.1.3, and an additional permit ACE to allow other reserved administrative IP addresses from 192.168.1.4 to 192.168.1.7. Notice how the first permit ACE signifies a single host, by using the **host** keyword, the ACE **permit 192.168.1.3 0.0.0.0** could have been used instead. The second permit ACE allows hosts 192.168.1.4 through 192.168.1.7, by using the 0.0.0.3 wildcard, which is the inverse of a 255.255.255.252 subnet mask.

```
R1(config-std-nacl)# permit host 192.168.1.3

R1(config-std-nacl)# permit 192.168.1.4 0.0.0.3

R1(config-std-nacl)# exit
```

You do not need to enter a deny ACE because there is an implicit **deny any** ACE at the end of the ACL.

g. Now that the named ACL is created, apply it to the vty lines.

```
R1(config)# line vty 0 4

R1(config-line)# access-class ADMIN-MGT in

R1(config-line)# exit
```

Part 3: Verify the Access Control List Using Telnet

In Part 3, you will use Telnet to access the router, verifying that the named ACL is functioning correctly.

Note: SSH is more secure than Telnet; however, SSH requires that the network device be configured to accept SSH connections. Telnet is used with this lab for convenience.

a. Open a command prompt on PC-A and verify that you can communicate with the router by issuing a **ping** command.

```
C:\Users\user1> ping 192.168.1.1


Pinging 192.168.1.1 with 32 bytes of data:
Reply from 192.168.1.1: bytes=32 time=5ms TTL=64
Reply from 192.168.1.1: bytes=32 time=1ms TTL=64
Reply from 192.168.1.1: bytes=32 time=1ms TTL=64
Reply from 192.168.1.1: bytes=32 time=1ms TTL=64


Ping statistics for 192.168.1.1:
    Packets: Sent = 4, Received = 4, Lost = 0 (0% loss),
Approximate round trip times in milli-seconds:
    Minimum = 1ms, Maximum = 5ms, Average = 2ms
C:\Users\user1>
```

b. Using the command prompt on PC-A, launch the Telnet client program to telnet into the router. Enter the login and then the enable passwords. You should be successfully logged in, see the banner message, and receive an R1 router command prompt.

```
C:\Users\user1> telnet 192.168.1.1


Unauthorized access is prohibited!


User Access Verification


Password:
R1>enable
Password:
R1#
```

Was the Telnet connection successful? _____

c. Type **exit** at the command prompt and press Enter to exit the Telnet session.

d. Change your IP address to test if the named ACL blocks non-permitted IP addresses. Change the IPv4 address to 192.168.1.100 on PC-A.

e. Attempt to telnet into R1 at 192.168.1.1 again. Was the Telnet session successful?

_____

What message was received? _____

f. Change the IP address on PC-A to test if the named ACL permits a host with an IP address from the 192.168.1.4 to 192.168.1.7 range to telnet into the router. After changing the IP address on PC-A, open a Windows command prompt and attempt to telnet into router R1.

Was the Telnet session successful?

_____

_____

g. From privileged EXEC mode on R1, type the **show ip access-lists** command and press Enter. From the command output, notice how the Cisco IOS automatically assigns line numbers to the ACL ACEs in increments of 10 and shows the number of times each permit ACE has been successfully matched (in parenthesis).

```
R1# show ip access-lists
Standard IP access list ADMIN-MGT
    10 permit 192.168.1.3 (2 matches)
    20 permit 192.168.1.4, wildcard bits 0.0.0.3 (2 matches)
```

Because two successful Telnet connections to the router were established, and each Telnet session was initiated from an IP address that matches one of the permit ACEs, there are matches for each permit ACE.

Why do you think that there are two matches for each permit ACE when only one connection from each IP address was initiated?

_____

_____

_____

_____

How would you determine at what point the Telnet protocol causes the two matches during the Telnet connection?

_____

_____

_____

h. On R1, enter into global configuration mode.

i. Enter into access-list configuration mode for the ADMIN-MGT named access list and add a **deny any** ACE to the end of the access list.

```
R1(config)# ip access-list standard ADMIN-MGT
R1(config-std-nacl)# deny any
R1(config-std-nacl)# exit
```

Note: Because there is an implicit **deny any** ACE at the end of all ACLs, adding an explicit **deny any** ACE is unnecessary, yet can still be useful to the network administrator to log or simply know how many times the **deny any** access-list ACE was matched.

j.   Try to telnet from PC-B to R1. This creates a match to the **deny any** ACE in the ADMIN-MGT named access list.

k.   From privileged EXEC mode, type **show ip access-lists** command and press Enter. You should now see multiple matches to the **deny any** ACE.

```
R1# show ip access-lists
Standard IP access list ADMIN-MGT
     10 permit 192.168.1.3 (2 matches)
     20 permit 192.168.1.4, wildcard bits 0.0.0.3 (2 matches)
     30 deny any (3 matches)
```

The failed Telnet connection produces more matches to the explicit deny ACE than a successful one. Why do you think this happens?

_____

_____

Part 4:   Challenge - Configure and Apply the Access Control List on S1

Step 1:   Configure and apply a standard named ACL for the vty lines on S1.

a.   Without referring back to the R1 configuration commands, try to configure the ACL on S1, allowing only the PC-A IP address.

b.   Apply the ACL to the S1 vty lines. Remember that there are more vty lines on a switch than a router.

Step 2:   Test the vty ACL on S1.

Telnet from each of the PCs to verify that the vty ACL is working properly. You should be able to telnet to S1 from PC-A, but not from PC-B.

Reflection

1.   As evidenced by the remote vty access, ACLs are powerful content filters that can be applied to more than just inbound and outbound network interfaces. It what other ways might ACLs be applied?

_____

_____

2.   Does an ACL applied to a vty remote management interface improve the security of Telnet connection? Does this make Telnet a more viable remote access management tool?

_____

_____

_____

3. Why does it make sense to apply an ACL to vty lines instead of specific interfaces?

_____

_____

_____

Router Interface Summary Table

Router Interface Summary				
Router Model	**Ethernet Interface #1**	**Ethernet Interface #2**	**Serial Interface #1**	**Serial Interface #2**
1800	Fast Ethernet 0/0 (F0/0)	Fast Ethernet 0/1 (F0/1)	Serial 0/0/0 (S0/0/0)	Serial 0/0/1 (S0/0/1)
1900	Gigabit Ethernet 0/0 (G0/0)	Gigabit Ethernet 0/1 (G0/1)	Serial 0/0/0 (S0/0/0)	Serial 0/0/1 (S0/0/1)
2801	Fast Ethernet 0/0 (F0/0)	Fast Ethernet 0/1 (F0/1)	Serial 0/1/0 (S0/1/0)	Serial 0/1/1 (S0/1/1)
2811	Fast Ethernet 0/0 (F0/0)	Fast Ethernet 0/1 (F0/1)	Serial 0/0/0 (S0/0/0)	Serial 0/0/1 (S0/0/1)
2900	Gigabit Ethernet 0/0 (G0/0)	Gigabit Ethernet 0/1 (G0/1)	Serial 0/0/0 (S0/0/0)	Serial 0/0/1 (S0/0/1)
Note: To find out how the router is configured, look at the interfaces to identify the type of router and how many interfaces the router has. There is no way to effectively list all the combinations of configurations for each router class. This table includes identifiers for the possible combinations of Ethernet and Serial interfaces in the device. The table does not include any other type of interface, even though a specific router may contain one. An example of this might be an ISDN BRI interface. The string in parenthesis is the legal abbreviation that can be used in Cisco IOS commands to represent the interface.				

9.3.2.13 Lab – Configuring and Verifying Extended ACLs

Topology

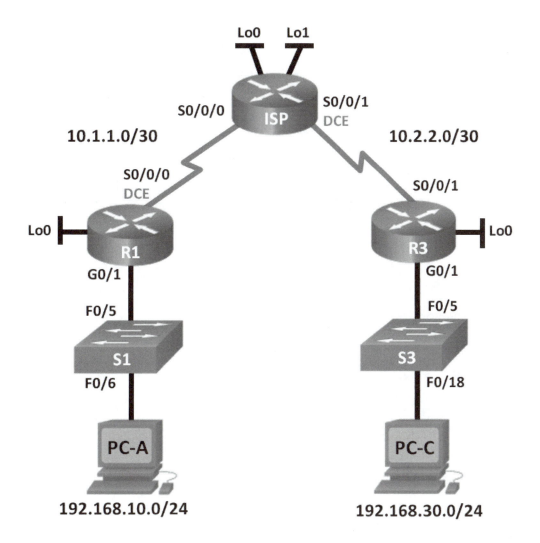

Addressing Table

Device	Interface	IP Address	Subnet Mask	Default Gateway
R1	G0/1	192.168.10.1	255.255.255.0	N/A
	Lo0	192.168.20.1	255.255.255.0	N/A
	S0/0/0 (DCE)	10.1.1.1	255.255.255.252	N/A
ISP	S0/0/0	10.1.1.2	255.255.255.252	N/A
	S0/0/1 (DCE)	10.2.2.2	255.255.255.252	N/A
	Lo0	209.165.200.225	255.255.255.224	N/A
	Lo1	209.165.201.1	255.255.255.224	N/A
R3	G0/1	192.168.30.1	255.255.255.0	N/A
	Lo0	192.168.40.1	255.255.255.0	N/A
	S0/0/1	10.2.2.1	255.255.255.252	N/A
S1	VLAN 1	192.168.10.11	255.255.255.0	192.168.10.1
S3	VLAN 1	192.168.30.11	255.255.255.0	192.168.30.1
PC-A	NIC	192.168.10.3	255.255.255.0	192.168.10.1
PC-C	NIC	192.168.30.3	255.255.255.0	192.168.30.1

Objectives

Part 1: Set Up the Topology and Initialize Devices

Part 2: Configure Devices and Verify Connectivity

- Configure basic settings on PCs, routers, and switches.
- Configure EIGRP routing on R1, ISP, and R3.

Part 3: Configure and Verify Extended Numbered and Named ACLs

- Configure, apply, and verify a numbered extended ACL.
- Configure, apply, and verify a named extended ACL.

Part 4: Modify and Verify Extended ACLs

Background / Scenario

Extended access control lists (ACLs) are extremely powerful. They offer a much greater degree of control than standard ACLs as to the types of traffic that can be filtered, as well as where the traffic originated and where it is going.

In this lab, you will set up filtering rules for two offices represented by R1 and R3. Management has established some access policies between the LANs located at R1 and R3, which you must implement. The ISP router between R1 and R3 does not have any ACLs placed on it. You would not be allowed any administrative access to an ISP router as you can only control and manage your own equipment.

Note: The routers used with CCNA hands-on labs are Cisco 1941 Integrated Services Routers (ISRs) with Cisco IOS Release 15.2(4)M3 (universalk9 image). The switches used are Cisco Catalyst 2960s with Cisco IOS Release 15.0(2) (lanbasek9 image). Other routers, switches, and Cisco IOS versions can be used. Depending on the model and Cisco IOS version, the commands available and output produced might vary from what is shown in the labs. Refer to the Router Interface Summary Table at the end of the lab for the correct interface identifiers.

Note: Make sure that the routers and switches have been erased and have no startup configurations. If you are unsure, contact your instructor.

Required Resources

- 3 Routers (Cisco 1941 with Cisco IOS Release 15.2(4)M3 universal image or comparable)
- 2 Switches (Cisco 2960 with Cisco IOS Release 15.0(2) lanbasek9 image or comparable)
- 2 PCs (Windows 7, Vista, or XP with terminal emulation program, such as Tera Term)
- Console cables to configure the Cisco IOS devices via the console ports
- Ethernet and serial cables as shown in the topology

Part 1: Set Up the Topology and Initialize Devices

In Part 1, you will set up the network topology and clear any configurations if necessary.

Step 1: Cable the network as shown in the topology.

Step 2: Initialize and reload the routers and switches.

Part 2: Configure Devices and Verify Connectivity

In Part 2, you will configure basic settings on the routers, switches, and PCs. Refer to the Topology and Addressing Table for device names and address information.

Step 1: Configure IP addresses on PC-A and PC-C.

Step 2: Configure basic settings on R1.

a. Disable DNS lookup.

b. Configure the device name as shown in the topology.

c. Create a loopback interface on R1.

d. Configure interface IP addresses as shown in the Topology and Addressing Table.

e. Configure a privileged EXEC mode password of **class**.

f. Assign a clock rate of **128000** to the S0/0/0 interface.

g. Assign **cisco** as the console and vty password and enable Telnet access. Configure **logging synchronous** for both the console and vty lines.

h. Enable web access on R1 to simulate a web server with local authentication for user **admin**.

```
R1(config)# ip http server
R1(config)# ip http authentication local
R1(config)# username admin privilege 15 secret class
```

Step 3: **Configure basic settings on ISP.**

    a.   Configure the device name as shown in the topology.

    b.   Create the loopback interfaces on ISP.

    c.   Configure interface IP addresses as shown in the Topology and Addressing Table.

    d.   Disable DNS lookup.

    e.   Assign **class** as the privileged EXEC mode password.

    f.   Assign a clock rate of **128000** to the S0/0/1 interface.

    g.   Assign **cisco** as the console and vty password and enable Telnet access. Configure **logging synchronous** for both console and vty lines.

    h.   Enable web access on the ISP. Use the same parameters as in Step 2h.

Step 4: **Configure basic settings on R3.**

    a.   Configure the device name as shown in the topology.

    b.   Create a loopback interface on R3.

    c.   Configure interface IP addresses as shown in the Topology and Addressing Table.

    d.   Disable DNS lookup.

    e.   Assign **class** as the privileged EXEC mode password.

    f.   Assign **cisco** as the console password and configure **logging synchronous** on the console line.

    g.   Enable SSH on R3.

```
R3(config)# ip domain-name cisco.com
R3(config)# crypto key generate rsa modulus 1024
R3(config)# line vty 0 4
R3(config-line)# login local
R3(config-line)# transport input ssh
```

    h.   Enable web access on R3. Use the same parameters as in Step 2h.

Step 5: **(Optional) Configure basic settings on S1 and S3.**

    a.   Configure the hostnames as shown in the topology.

    b.   Configure the management interface IP addresses as shown in the Topology and Addressing Table.

    c.   Disable DNS lookup.

    d.   Configure a privileged EXEC mode password of **class**.

    e.   Configure a default gateway address.

Step 6: **Configure EIGRP routing on R1, ISP, and R3.**

    a.   Configure autonomous system (AS) number 10 and advertise all networks on R1, ISP, and R3. Disable automatic summarization.

    b.   After configuring EIGRP on R1, ISP, and R3, verify that all routers have complete routing tables listing all networks. Troubleshoot if this is not the case.

Step 7: **Verify connectivity between devices.**

    Note: It is very important to verify connectivity **before** you configure and apply ACLs! Ensure that your network is properly functioning before you start to filter out traffic.

    a.   From PC-A, ping PC-C and the loopback and serial interfaces on R3.

        Were your pings successful?

    b.   From R1, ping PC-C and the loopback and serial interface on R3.

        Were your pings successful? _____

    c.   From PC-C, ping PC-A and the loopback and serial interface on R1.

        Were your pings successful? _____

    d.   From R3, ping PC-A and the loopback and serial interface on R1.

        Were your pings successful?

    e.   From PC-A, ping the loopback interfaces on the ISP router.

        Were your pings successful? _____

    f.   From PC-C, ping the loopback interfaces on the ISP router.

        Were your pings successful? _____

    g.   Open a web browser on PC-A and go to http://209.165.200.225 on ISP. You will be prompted for a username and password. Use **admin** for the username and **class** for the password. If you are prompted to accept a signature, accept it. The router will load the Cisco Configuration Professional (CCP) Express in a separate window. You may be prompted for a username and password. Use **admin** for the username and **class** for the password.

    h.   Open a web browser on PC-C and go to http://10.1.1.1 on R1. You will be prompted for a username and password. Use **admin** for username and **class** for the password. If you are prompted to accept a signature, accept it. The router will load CCP Express in a separate window. You may be prompted for a username and password. Use **admin** for the username and **class** for the password.

Part 3: **Configure and Verify Extended Numbered and Named ACLs**

Extended ACLs can filter traffic in many different ways. Extended ACLs can filter on source IP addresses, source ports, destination IP addresses, destination ports, as well as various protocols and services.

Security policies are as follows:

1.   Allow web traffic originating from the 192.168.10.0/24 network to go to any network.

2.   Allow an SSH connection to the R3 serial interface from PC-A.

3.   Allow users on 192.168.10.0.24 network access to 192.168.20.0/24 network.

4.   Allow web traffic originating from the 192.168.30.0/24 network to access R1 via the web interface and the 209.165.200.224/27 network on ISP. The 192.168.30.0/24 network should NOT be allowed to access any other network via the web.

In looking at the security policies listed above, you will need at least two ACLs to fulfill the security policies. A best practice is to place extended ACLs as close to the source as possible. We will follow this practice for these policies.

Step 1: **Configure a numbered extended ACL on R1 for security policy numbers 1 and 2.**

You will use a numbered extended ACL on R1. What are the ranges for extended ACLs?

_____

a. Configure the ACL on R1. Use 100 for the ACL number.

```
R1(config)# access-list 100 remark Allow Web & SSH Access
R1(config)# access-list 100 permit tcp host 192.168.10.3 host 10.2.2.1 eq 22
R1(config)# access-list 100 permit tcp any any eq 80
```

What does the 80 signify in the command output listed above?

_____

To what interface should ACL 100 be applied?

_____

_____

In what direction should ACL 100 be applied?

_____

_____

b. Apply ACL 100 to the S0/0/0 interface.

```
R1(config)# int s0/0/0
R1(config-if)# ip access-group 100 out
```

c. Verify ACL 100.

1) Open up a web browser on PC-A, and access http://209.165.200.225 (the ISP router). It should be successful; troubleshoot, if not.

2) Establish an SSH connection from PC-A to R3 using 10.2.2.1 for the IP address. Log in with **admin** and **class** for your credentials. It should be successful; troubleshoot, if not.

3) From privileged EXEC mode prompt on R1, issue the **show access-lists** command.

```
R1# show access-lists
Extended IP access list 100
    10 permit tcp host 192.168.10.3 host 10.2.2.1 eq 22 (22 matches)
    20 permit tcp any any eq www (111 matches)
```

4) From the PC-A command prompt, issue a ping to 10.2.2.1. Explain your results?

_____

_____

_____

Step 2: Configure a named extended ACL on R3 for security policy number 3.

a. Configure the policy on R3. Name the ACL WEB-POLICY.

R3(config)# **ip access-list extended WEB-POLICY**

R3(config-ext-nacl)# **permit tcp 192.168.30.0 0.0.0.255 host 10.1.1.1 eq 80**

R3(config-ext-nacl)# **permit tcp 192.168.30.0 0.0.0.255 209.165.200.224 0.0.0.31 eq 80**

b. Apply ACL WEB-POLICY to the S0/0/1 interface.

R3(config-ext-nacl)# **int S0/0/1**

R3(config-if)# **ip access-group WEB-POLICY out**

c. Verify the ACL WEB-POLICY.

1) From R3 privileged EXEC mode command prompt, issue the **show ip interface s0/0/1** command.

What, if any, is the name of the ACL? _____

In what direction is the ACL applied? _____

2) Open up a web browser on PC-C and access http://209.165.200.225 (the ISP router). It should be successful; troubleshoot, if not.

3) From PC-C, open a web session to http://10.1.1.1 (R1). It should be successful; troubleshoot, if not.

4) From PC-C, open a web session to http://209.165.201.1 (ISP router). It should fail; troubleshoot, if not.

5) From a PC-C command prompt, ping PC-A. What was your result and why?

_____

_____

Part 4: **Modify and Verify Extended ACLs**

Because of the ACLs applied on R1 and R3, no pings or any other kind of traffic is allowed between the LAN networks on R1 and R3. Management has decided that all traffic between the 192.168.10.0/24 and 192.168.30.0/24 networks should be allowed. You must modify both ACLs on R1 and R3.

Step 1: **Modify ACL 100 on R1.**

a. From R1 privileged EXEC mode, issue the **show access-lists** command.

How many lines are there in this access list? _____

b. Enter global configuration mode and modify the ACL on R1.

```
R1(config)# ip access-list extended 100
R1(config-ext-nacl)# 30 permit ip 192.168.10.0 0.0.0.255 192.168.30.0 0.0.0.255
R1(config-ext-nacl)# end
```

c. Issue the **show access-lists** command.

Where did the new line that you just added appear in ACL 100?

_____

Step 2: **Modify ACL WEB-POLICY on R3.**

a. From R3 privileged EXEC mode, issue the **show access-lists** command.

How many lines are there in this access list? _____

b. Enter global configuration mode and modify the ACL on R3.

```
R3(config)# ip access-list extended WEB-POLICY
R3(config-ext-nacl)# 30 permit ip 192.168.30.0 0.0.0.255 192.168.10.0 0.0.0.255
R3(config-ext-nacl)# end
```

c. Issue the **show access-lists** command to verify that the new line was added at the end of the ACL.

Step 3: **Verify modified ACLs.**

a. From PC-A, ping the IP address of PC-C. Were the pings successful? _____

b. From PC-C, ping the IP address of PC-A. Were the pings successful? _____

Why did the ACLs work immediately for the pings after you changed them?

_____

Reflection

1. Why is careful planning and testing of ACLs required?

_____

_____

2. Which type of ACL is better: standard or extended?

_____

_____

_____

_____

3. Why are EIGRP hello packets and routing updates not blocked by the implicit **deny any** access control entry (ACE) or ACL statement of the ACLs applied to R1 and R3?

_____

_____

_____

_____

Router Interface Summary Table

Router Interface Summary				
Router Model	**Ethernet Interface #1**	**Ethernet Interface #2**	**Serial Interface #1**	**Serial Interface #2**
1800	Fast Ethernet 0/0 (F0/0)	Fast Ethernet 0/1 (F0/1)	Serial 0/0/0 (S0/0/0)	Serial 0/0/1 (S0/0/1)
1900	Gigabit Ethernet 0/0 (G0/0)	Gigabit Ethernet 0/1 (G0/1)	Serial 0/0/0 (S0/0/0)	Serial 0/0/1 (S0/0/1)
2801	Fast Ethernet 0/0 (F0/0)	Fast Ethernet 0/1 (F0/1)	Serial 0/1/0 (S0/1/0)	Serial 0/1/1 (S0/1/1)
2811	Fast Ethernet 0/0 (F0/0)	Fast Ethernet 0/1 (F0/1)	Serial 0/0/0 (S0/0/0)	Serial 0/0/1 (S0/0/1)
2900	Gigabit Ethernet 0/0 (G0/0)	Gigabit Ethernet 0/1 (G0/1)	Serial 0/0/0 (S0/0/0)	Serial 0/0/1 (S0/0/1)
Note: To find out how the router is configured, look at the interfaces to identify the type of router and how many interfaces the router has. There is no way to effectively list all the combinations of configurations for each router class. This table includes identifiers for the possible combinations of Ethernet and Serial interfaces in the device. The table does not include any other type of interface, even though a specific router may contain one. An example of this might be an ISDN BRI interface. The string in parenthesis is the legal abbreviation that can be used in Cisco IOS commands to represent the interface.				

9.4.2.7 Lab - Troubleshooting ACL Configuration and Placement

Topology

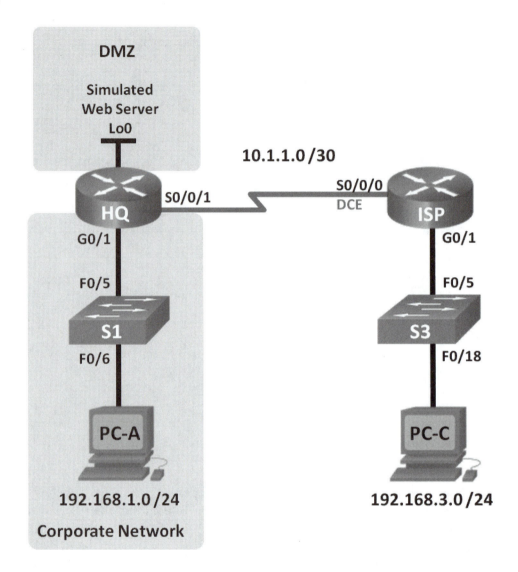

Addressing Table

Device	Interface	IP Address	Subnet Mask	Default Gateway
HQ	G0/1	192.168.1.1	255.255.255.0	N/A
	S0/0/1	10.1.1.2	255.255.255.252	N/A
	Lo0	192.168.4.1	255.255.255.0	N/A
ISP	G0/1	192.168.3.1	255.255.255.0	N/A
	S0/0/0 (DCE)	10.1.1.1	255.255.255.252	N/A
S1	VLAN 1	192.168.1.11	255.255.255.0	192.168.1.1
S3	VLAN 1	192.168.3.11	255.255.255.0	192.168.3.1
PC-A	NIC	192.168.1.3	255.255.255.0	192.168.1.1
PC-C	NIC	192.168.3.3	255.255.255.0	192.168.3.1

Objectives

Part 1: Build the Network and Configure Basic Device Settings

Part 2: Troubleshoot Internal Access

Part 3: Troubleshoot Remote Access

Background / Scenario

An access control list (ACL) is a series of IOS commands that can provide basic traffic filtering on a Cisco router. ACLs are used to select the types of traffic to be processed. A single ACL statement is called and access control entry (ACE). The ACEs in the ACL are evaluated from top to bottom with an implicit deny all ACE at the end of the list. ACLs can also control the types of traffic into or out of a network by the source and destination hosts or network. To process the desired traffic correctly, the placement of the ACLs is critical.

In this lab, a small company has just added a web server to the network to allow customers to access confidential information. The company network is divided into two zones: Corporate network zone and Demilitarized Zone (DMZ). The corporate network zone houses private servers and internal clients. The DMZ houses the externally accessible web server (simulated by Lo0 on HQ). Because the company can only administer its own HQ router, all ACLs must be applied to the HQ router.

- ACL 101 is implemented to limit the traffic out of the corporate network zone. This zone houses the private servers and internal clients (192.168.1.0/24). No other network should be able to access it.

- ACL 102 is used to limit the traffic into the corporate network. Only responses to requests that originated from within the corporate network are allowed back into that network. This includes TCP-based requests from internal hosts such as Web and FTP. ICMP is allowed into the network for troubleshooting purposes so that incoming ICMP messages generated in response to pings can be received by internal hosts.

- ACL 121 controls outside traffic to the DMZ and corporate network. Only HTTP traffic is allowed to the DMZ web server (simulated by Lo0 on R1). Other network related traffic, such as EIGRP, is allowed from outside networks. Furthermore, valid internal private addresses, such as 192.168.1.0, loopback address such as 127.0.0.0 and multicast addresses are denied entrance to the corporate network to prevent malicious network attacks from outside users.

Note: The routers used with CCNA hands-on labs are Cisco 1941 Integrated Services Routers (ISRs) with Cisco IOS Release 15.2(4)M3 (universalk9 image). The switches used are Cisco Catalyst 2960s with Cisco IOS Release 15.0(2) (lanbasek9 image). Other routers, switches and Cisco IOS versions can be used. Depending on the model and Cisco IOS version, the commands available and output produced might vary from what is shown in the labs. Refer to the Router Interface Summary Table at the end of the lab for the correct interface identifiers.

Note: Make sure that the routers and switches have been erased and have no startup configurations. If you are unsure, contact your instructor.

Required Resources

- 2 Routers (Cisco 1941 with Cisco IOS Release 15.2(4)M3 universal image or comparable)

- 2 Switches (Cisco 2960 with Cisco IOS Release 15.0(2) lanbasek9 image or comparable)

- 2 PCs (Windows 7, Vista, or XP with terminal emulation program, such as Tera Term)

- Console cables to configure the Cisco IOS devices via the console ports

- Ethernet and serial cables as shown in the topology

Part 1:   Build the Network and Configure Basic Device Settings

In Part 1, you set up the network topology and configure the routers and switches with some basic settings, such as passwords and IP addresses. Preset configurations are also provided for you for the initial router configurations. You will also configure the IP settings for the PCs in the topology.

Step 1:   Cable the network as shown in the topology.

Step 2:   Configure PC hosts.

Step 3:   Initialize and reload the routers and switches as necessary.

Step 4:   (Optional) Configure basic settings for each switch.

a.   Disable DNS lookup.

b.   Configure host names as shown in the Topology.

c.   Configure IP address and default gateway in Addressing Table.

d.   Assign **cisco** as the console and vty passwords.

e.   Assign **class** as the privileged EXEC password.

f.   Configure **logging synchronous** to prevent console messages from interrupting command entry.

Step 5:   Configure basic settings for each router.

a.   Disable DNS lookup.

b.   Configure host names as shown in the topology.

c.   Assign **cisco** as the console and vty passwords.

d.   Assign **class** as the privileged EXEC password.

e.   Configure **logging synchronous** to prevent console messages from interrupting command entry.

Step 6:   Configure HTTP access and user credentials on HQ router.

Local user credentials are configured to access the simulated web server (192.168.4.1).

```
HQ(config)# ip http server
HQ(config)# username admin privilege 15 secret adminpass
HQ(config)# ip http authentication local
```

Step 7:   Load router configurations.

The configurations for the routers ISP and HQ are provided for you. There are errors within these configurations, and it is your job to determine the incorrect configurations and correct them.

Router ISP

```
hostname ISP
interface GigabitEthernet0/1
 ip address 192.168.3.1 255.255.255.0
 no shutdown
interface Serial0/0/0
```

```
 ip address 10.1.1.1 255.255.255.252
 clock rate 128000
 no shutdown
router eigrp 1
 network 10.1.1.0 0.0.0.3
 network 192.168.3.0
 no auto-summary
end
```

Router HQ

```
hostname HQ
interface Loopback0
 ip address 192.168.4.1 255.255.255.0
interface GigabitEthernet0/1
 ip address 192.168.1.1 255.255.255.0
 ip access-group 101 out


 ip access-group 102 in


 no shutdown
interface Serial0/0/1
 ip address 10.1.1.2 255.255.255.252
 ip access-group 121 in
 no shutdown
router eigrp 1
 network 10.1.1.0 0.0.0.3
 network 192.168.1.0
 network 192.168.4.0
 no auto-summary
access-list 101 permit ip 192.168.11.0 0.0.0.255 any

access-list 101 deny ip any any
access-list 102 permit tcp any any established
access-list 102 permit icmp any any echo-reply
access-list 102 permit icmp any any unreachable
access-list 102 deny ip any any
access-list 121 permit tcp any host 192.168.4.1 eq 89

access-list 121 deny icmp any host 192.168.4.11
```

```
access-list 121 deny ip 192.168.1.0 0.0.0.255 any

access-list 121 deny ip 127.0.0.0 0.255.255.255 any

access-list 121 deny ip 224.0.0.0 31.255.255.255 any

access-list 121 permit ip any any

access-list 121 deny ip any any

end
```

Part 2: Troubleshoot Internal Access

In Part 2, the ACLs on router HQ are examined to determine if they are configured correctly.

Step 1: Troubleshoot ACL 101

ACL 101 is implemented to limit the traffic out of the corporate network zone. This zone houses only internal clients and private servers. Only 192.168.1.0/24 network can exit this corporate network zone.

a.   Can PC-A ping its default gateway? _____

b.   After verifying that the PC-A was configured correctly, examine the HQ router to find possible configuration errors by viewing the summary of ACL 101. Enter the command **show access-lists 101**.

```
HQ# show access-lists 101

Extended IP access list 101

    10 permit ip 192.168.11.0 0.0.0.255 any

    20 deny ip any any
```

c.   Are there any problems with ACL 101?

---

d.   Examine the default gateway interface for the 192.168.1.0 /24 network. Verify that the ACL 101 is applied in the correct direction on the G0/1 interface. Enter the **show ip interface g0/1** command.

```
HQ# show ip interface g0/1

GigabitEthernet0/1 is up, line protocol is up

   Internet address is 192.168.1.1/24

   Broadcast address is 255.255.255.255

   Address determined by setup command

   MTU is 1500 bytes

   Helper address is not set

   Directed broadcast forwarding is disabled

   Multicast reserved groups joined: 224.0.0.10

   Outgoing access list is 101

   Inbound  access list is 102
```

Is the direction for interface G0/1 configured correctly for ACL 101?

---

e. Correct the errors found regarding ACL 101 and verify the traffic from network 192.168.1.0 /24 can exit the corporate network. Record the commands used to correct the errors.

_____

_____

_____

_____

_____

f. Verify PC-A can ping its default gateway interface.

Step 2: **Troubleshoot ACL 102**

ACL 102 is implemented to limit traffic into the corporate network. Traffic originating from the outside network is not allowed onto the corporate network. Remote traffic is allowed into the corporate network if the established traffic originated from the internal network. ICMP reply messages are allowed for troubleshooting purposes.

a. Can PC-A ping PC-C? _____

b. Examine the HQ router to find possible configuration errors by viewing the summary of ACL 102. Enter the command **show access-lists 102**.

```
HQ# show access-lists 102
Extended IP access list 102
    10 permit tcp any any established
    20 permit icmp any any echo-reply
    30 permit icmp any any unreachable
    40 deny ip any any (57 matches)
```

c. Are there any problems with ACL 102?

_____

d. Verify that the ACL 102 is applied in the correct direction on G0/1 interface. Enter the **show ip interface g0/1** command.

```
HQ# show ip interface g0/1
GigabitEthernet0/1 is up, line protocol is up
  Internet address is 192.168.1.1/24
  Broadcast address is 255.255.255.255
  Address determined by setup command
```

```
MTU is 1500 bytes

Helper address is not set

Directed broadcast forwarding is disabled

Multicast reserved groups joined: 224.0.0.10

Outgoing access list is 101

Inbound  access list is 101
```

e. Are there any problems with the application of ACL 102 to interface G0/1?

_____

f. Correct any errors found regarding ACL 102. Record the commands used to correct the errors.

g. Can PC-A ping PC-C now? _____

Part 3: **Troubleshoot Remote Access**

In Part 3, ACL 121 is configured to prevent spoofing attacks from the outside networks and allow only remote HTTP access to the web server (192.168.4.1) in DMZ.

a. Verify ACL 121 has been configured correctly. Enter the **show ip access-list 121** command.

```
HQ# show ip access-lists 121

Extended IP access list 121

     10 permit tcp any host 192.168.4.1 eq 89

     20 deny icmp any host 192.168.4.11

     30 deny ip 192.168.1.0 0.0.0.255 any

     40 deny ip 127.0.0.0 0.255.255.255 any

     50 deny ip 224.0.0.0 31.255.255.255 any

     60 permit ip any any (354 matches)

     70 deny ip any any
```

Are there any problems with this ACL?

_____

_____

b. Verify that the ACL 121 is applied in the correct direction on the R1 S0/0/1 interface. Enter the **show ip interface s0/0/1** command.

```
HQ# show ip interface s0/0/1

Serial0/0/1 is up, line protocol is up

  Internet address is 10.1.1.2/30

  Broadcast address is 255.255.255.255

<output omitted>

  Multicast reserved groups joined: 224.0.0.10

  Outgoing access list is not set

  Inbound  access list is 121
```

Are there any problems with the application of this ACL?

_____

_____

c. If any errors were found, make and record the necessary configuration changes to ACL 121.

_____

_____

_____

_____

_____

_____

d. Verify that PC-C can only access the simulated web server on HQ by using the web browser. Provide the username **admin** and password **adminpass** to access the web server (192.168.4.1).

Reflection

1. How should the ACL statement be ordered? From general to specific or vice versa?

_____

2. If you delete an ACL by using the **no access-list** command and the ACL is still applied to the interface, what happens?

_____

Router Interface Summary Table

Router Interface Summary				
Router Model	**Ethernet Interface #1**	**Ethernet Interface #2**	**Serial Interface #1**	**Serial Interface #2**
1800	Fast Ethernet 0/0 (F0/0)	Fast Ethernet 0/1 (F0/1)	Serial 0/0/0 (S0/0/0)	Serial 0/0/1 (S0/0/1)
1900	Gigabit Ethernet 0/0 (G0/0)	Gigabit Ethernet 0/1 (G0/1)	Serial 0/0/0 (S0/0/0)	Serial 0/0/1 (S0/0/1)
2801	Fast Ethernet 0/0 (F0/0)	Fast Ethernet 0/1 (F0/1)	Serial 0/1/0 (S0/1/0)	Serial 0/1/1 (S0/1/1)
2811	Fast Ethernet 0/0 (F0/0)	Fast Ethernet 0/1 (F0/1)	Serial 0/0/0 (S0/0/0)	Serial 0/0/1 (S0/0/1)
2900	Gigabit Ethernet 0/0 (G0/0)	Gigabit Ethernet 0/1 (G0/1)	Serial 0/0/0 (S0/0/0)	Serial 0/0/1 (S0/0/1)
Note: To find out how the router is configured, look at the interfaces to identify the type of router and how many interfaces the router has. There is no way to effectively list all the combinations of configurations for each router class. This table includes identifiers for the possible combinations of Ethernet and Serial interfaces in the device. The table does not include any other type of interface, even though a specific router may contain one. An example of this might be an ISDN BRI interface. The string in parenthesis is the legal abbreviation that can be used in Cisco IOS commands to represent the interface.				

9.5.2.7 Lab – Configuring and Verifying IPv6 ACLs

Topology

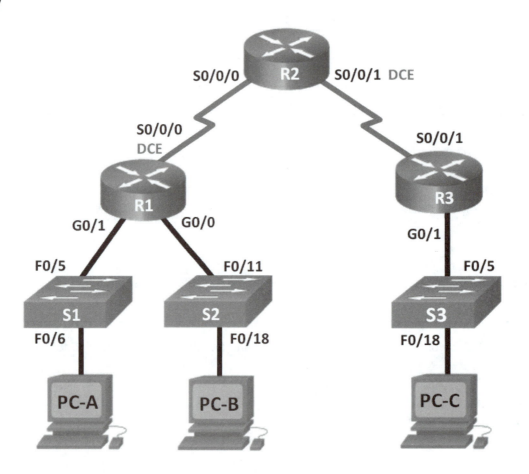

Addressing Table

Device	Interface	IP Address	Default Gateway
R1	G0/0	2001:DB8:ACAD:B::1/64	N/A
	G0/1	2001:DB8:ACAD:A::1/64	N/A
	S0/0/0 (DCE)	2001:DB8:AAAA:1::1/64	N/A
R2	S0/0/0	2001:DB8:AAAA:1::2/64	N/A
	S0/0/1 (DCE)	2001:DB8:AAAA:2::2/64	N/A
R3	G0/1	2001:DB8:CAFE:C::1/64	N/A
	S0/0/1	2001:DB8:AAAA:2::1/64	N/A
S1	VLAN1	2001:DB8:ACAD:A::A/64	N/A
S2	VLAN1	2001:DB8:ACAD:B::A/64	N/A
S3	VLAN1	2001:DB8:CAFE:C::A/64	N/A
PC-A	NIC	2001:DB8:ACAD:A::3/64	FE80::1
PC-B	NIC	2001:DB8:ACAD:B::3/64	FE80::1
PC-C	NIC	2001:DB8:CAFE:C::3/64	FE80::1

Objectives

Part 1: Set Up the Topology and Initialize Devices

Part 2: Configure Devices and Verify Connectivity

Part 3: Configure and Verify IPv6 ACLs

Part 4: Edit IPv6 ACLs

Background / Scenario

You can filter IPv6 traffic by creating IPv6 access control lists (ACLs) and applying them to interfaces similarly to the way that you create IPv4 named ACLs. IPv6 ACL types are extended and named. Standard and numbered ACLs are no longer used with IPv6. To apply an IPv6 ACL to a vty interface, you use the new **ipv6 traffic-filter** command. The **ipv6 access-class** command is still used to apply an IPv6 ACL to interfaces.

In this lab, you will apply IPv6 filtering rules and then verify that they are restricting access as expected. You will also edit an IPv6 ACL and clear the match counters.

Note: The routers used with CCNA hands-on labs are Cisco 1941 Integrated Services Routers (ISRs) with Cisco IOS Release 15.2(4)M3 (universalk9 image). The switches used are Cisco Catalyst 2960s with Cisco IOS Release 15.0(2) (lanbasek9 image). Other routers, switches and Cisco IOS versions can be used. Depending on the model and Cisco IOS version, the commands available and output produced might vary from what is shown in the labs. Refer to the Router Interface Summary Table at the end of the lab for the correct interface identifiers.

Note: Make sure that the routers and switches have been erased and have no startup configurations. If you are unsure, contact your instructor.

Required Resources

- 3 Routers (Cisco 1941 with Cisco IOS Release 15.2(4)M3 universal image or comparable)
- 3 Switches (Cisco 2960 with Cisco IOS Release 15.0(2) lanbasek9 image or comparable)
- 3 PCs (Windows 7, Vista, or XP with terminal emulation program, such as Tera Term)
- Console cables to configure the Cisco IOS devices via the console ports
- Ethernet and serial cables as shown in the topology

Part 1: Set Up the Topology and Initialize Devices

In Part 1, you set up the network topology and clear any configurations if necessary.

Step 1: Cable the network as shown in the topology.

Step 2: Initialize and reload the routers and switches.

Part 2: Configure Devices and Verify Connectivity

In Part 2, you configure basic settings on the routers, switches and PCs. Refer to the Topology and Addressing Table at the beginning of this lab for device names and address information.

Step 1: **Configure IPv6 addresses on all PCs.**

Configure IPv6 global unicast addresses according to the Addressing Table. Use the link-local address of **FE80::1** for the default-gateway on all PCs.

Step 2: **Configure the switches.**

a. Disable DNS lookup.

b. Assign the hostname.

c. Assign a domain-name of **ccna-lab.com**.

d. Encrypt plain text passwords.

e. Create a MOTD banner warning users that unauthorized access is prohibited.

f. Create a local user database with a username of **admin** and password as **classadm**.

g. Assign **class** as the privileged EXEC encrypted password.

h. Assign **cisco** as the console password and enable login.

i. Enable login on the VTY lines using the local database.

j. Generate a crypto rsa key for ssh using a modulus size of 1024 bits.

k. Change the transport input VTY lines to all for SSH and Telnet only.

l. Assign an IPv6 address to VLAN 1 according to the Addressing Table.

m. Administratively disable all inactive interfaces.

Step 3: **Configure basic settings on all routers.**

a. Disable DNS lookup.

b. Assign the hostname.

c. Assign a domain-name of **ccna-lab.com**.

d. Encrypt plain text passwords.

e. Create a MOTD banner warning users that unauthorized access is prohibited.

f. Create a local user database with a username of **admin** and password as **classadm**.

g. Assign **class** as the privileged EXEC encrypted password.

h. Assign **cisco** as the console password and enable login.

i. Enable login on the VTY lines using the local database.

j. Generate a crypto rsa key for ssh using a modulus size of 1024 bits.

k. Change the transport input VTY lines to all for SSH and Telnet only.

Step 4: **Configure IPv6 settings on R1.**

a. Configure the IPv6 unicast address on interface G0/0, G0/1, and S0/0/0.

b. Configure the IPv6 link-local address on interface G0/0, G0/1, and S0/0/0. Use **FE80::1** for the link-local address on all three interfaces.

c. Set the clock rate on S0/0/0 to 128000.

d. Enable the interfaces.

e. Enable IPv6 unicast routing.

f. Configure an IPv6 default route to use interface S0/0/0.

```
R1(config)# ipv6 route ::/0 s0/0/0
```

Step 5: **Configure IPv6 settings on R2.**

a. Configure the IPv6 unicast address on interface S0/0/0 and S0/0/1.

b. Configure the IPv6 link-local address on interface S0/0/0 and S0/0/1. Use **FE80::2** for the link-local address on both interfaces.

c. Set the clock rate on S0/0/1 to 128000.

d. Enable the interfaces.

e. Enable IPv6 unicast routing.

f. Configure static IPv6 routes for traffic handling of R1 and R3 LAN subnets.

```
R2(config)# ipv6 route 2001:db8:acad::/48 s0/0/0
R2(config)# ipv6 route 2001:db8:cafe:c::/64 s0/0/1
```

Step 6: **Configure IPv6 settings on R3.**

a. Configure the IPv6 unicast address on interface G0/1 and S0/0/1.

b. Configure the IPv6 link-local address on interface G0/1 and S0/0/1. Use **FE80::1** for the link-local address on both interfaces.

c. Enable the interfaces.

d. Enable IPv6 unicast routing.

e. Configure an IPv6 default route to use interface S0/0/1.

```
R3(config)# ipv6 route ::/0 s0/0/1
```

Step 7: **Verify connectivity.**

a. Each PC should be able to ping the other PCs in the topology.

b. Telnet to R1 from all PCs in the Topology.

c. SSH to R1 from all PCs in the Topology.

d. Telnet to S1 from all PCs in the Topology.

e. SSH to S1 from all PCs in the Topology.

f. Troubleshoot connectivity issues now because the ACLs that you create in Part 3 of this lab will restrict access to some areas of the network.

Note: Tera Term requires the target IPv6 address to be enclosed in brackets. Enter the IPv6 address as shown, click **OK** and then click **Continue** to accept the security warning and connect to the router.

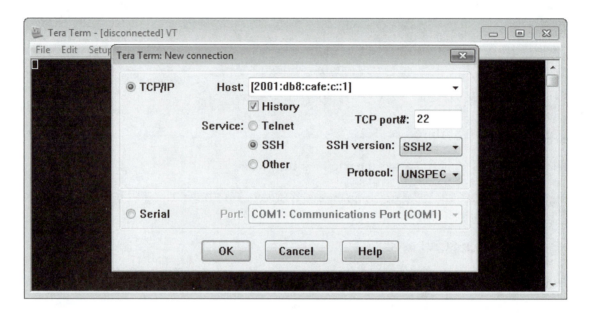

Input the user credentials configured (username **admin** and password **classadm**) and select the **Use plain password to log in** in the SSH Authentication dialogue box. Click **OK** to continue.

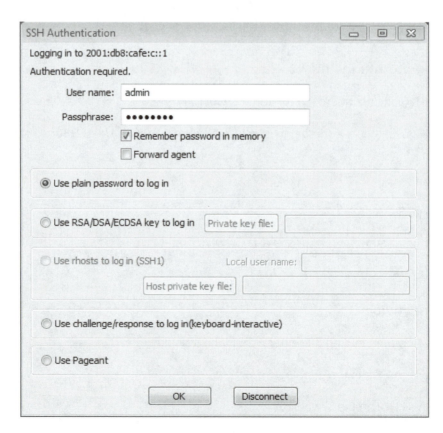

Part 3:   Configure and Verify IPv6 ACLs

Step 1:   Configure and verify VTY restrictions on R1.

a.   Create an ACL to only allow hosts from the 2001:db8:acad:a::/64 network to telnet to R1. All hosts should only be able to ssh to R1.

```
R1(config)# ipv6 access-list RESTRICT-VTY

R1(config-ipv6-acl)# permit tcp 2001:db8:acad:a::/64 any

R1(config-ipv6-acl)# permit tcp any any eq 22
```

b.   Apply the RESTRICT-VTY ACL to R1's VTY lines.

```
R1(config-ipv6-acl)# line vty 0 4

R1(config-line)# ipv6 access-class RESTRICT-VTY in

R1(config-line)# end

R1#
```

c.   Show the new ACL.

```
R1# show access-lists

IPv6 access list RESTRICT-VTY

    permit tcp 2001:DB8:ACAD:A::/64 any sequence 10

    permit tcp any any eq 22 sequence 20
```

d.   Verify that the RESTRICT-VTY ACL is only allowing Telnet traffic from the 2001:db8:acad:a::/64 network.

How does the RESTRICT-VTY ACL only allow hosts from the 2001:db8:acad:a::/64 network to telnet to R1?

_____

_____

_____

_____

What does the second permit statement in the RESTRICT-VTY ACL do?

_____

Step 2:   Restrict Telnet access to the 2001:db8:acad:a::/64 network.

a.   Create an ACL called RESTRICTED-LAN that will block Telnet access to the 2001:db8:acad:a::/64 network.

```
R1(config)# ipv6 access-list RESTRICTED-LAN

R1(config-ipv6-acl)# remark Block Telnet from outside

R1(config-ipv6-acl)# deny tcp any 2001:db8:acad:a::/64 eq telnet

R1(config-ipv6-acl)# permit ipv6 any any
```

b. Apply the RESTRICTED-LAN ACL to interface G0/1 for all outbound traffic.

```
R1(config-ipv6-acl)# int g0/1

R1(config-if)# ipv6 traffic-filter RESTRICTED-LAN out

R1(config-if)# end
```

c. Telnet to S1 from PC-B and PC-C to verify that Telnet has been restricted. SSH to S1 from PC-B to verify that it can still be reached using SSH. Troubleshoot if necessary.

d. Use the **show ipv6 access-list** command to view the RESTRICTED-LAN ACL.

```
R1# show ipv6 access-lists RESTRICTED-LAN

IPv6 access list RESTRICTED-LAN

    deny tcp any 2001:DB8:ACAD:A::/64 eq telnet (6 matches) sequence 20

    permit ipv6 any any (45 matches) sequence 30
```

Notice that each statement identifies the number of hits or matches that have occurred since the ACL was applied to the interface.

e. Use the **clear ipv6 access-list** to reset the match counters for the RESRICTED-LAN ACL.

```
R1# clear ipv6 access-list RESTRICTED-LAN
```

f. Redisplay the ACL with the **show access-lists** command to confirm that the counters were cleared.

```
R1# show access-lists RESTRICTED-LAN

IPv6 access list RESTRICTED-LAN

    deny tcp any 2001:DB8:ACAD:A::/64 eq telnet sequence 20

    permit ipv6 any any sequence 30
```

Part 4: Edit IPv6 ACLs

In Part 4, you will edit the RESTRICTED-LAN ACL that you created in Part 3. It is always a good idea to remove the ACL from the interface to which it is applied before editing it. After you complete your edits, then reapply the ACL to the interface.

Note: Many network administrators will make a copy of the ACL and edit the copy. When editing is complete, the administrator will remove the old ACL and apply the newly edited ACL to the interface. This method keeps the ACL in place until you are ready to apply the edited copy of the ACL.

Step 1: **Remove the ACL from the interface.**

```
R1(config)# int g0/1

R1(config-if)# no ipv6 traffic-filter RESTRICTED-LAN out

R1(config-if)# end
```

Step 2: **Use the show access-lists command to view the ACL.**

```
R1# show access-lists

IPv6 access list RESTRICT-VTY

    permit tcp 2001:DB8:ACAD:A::/64 any (4 matches) sequence 10

    permit tcp any any eq 22 (6 matches) sequence 20

IPv6 access list RESTRICTED-LAN
```

```
deny tcp any 2001:DB8:ACAD:A::/64 eq telnet sequence 20

permit ipv6 any any (36 matches) sequence 30
```

Step 3: Insert a new ACL statement using sequence numbering.

```
R1(config)# ipv6 access-list RESTRICTED-LAN

R1(config-ipv6-acl)# permit tcp 2001:db8:acad:b::/64 host 2001:db8:acad:a::a eq
23 sequence 15
```

What does this new permit statement do?

_____

Step 4: Insert a new ACL statement at the end of the ACL.

```
R1(config-ipv6-acl)# permit tcp any host 2001:db8:acad:a::3 eq www
```

Note: This permit statement is only used to show how to add a statement to the end of an ACL. This ACL line would never be matched because the previous permit statement is matching on everything.

Step 5: Use the do show access-lists command to view the ACL change.

```
R1(config-ipv6-acl)# do show access-list
IPv6 access list RESTRICT-VTY
    permit tcp 2001:DB8:ACAD:A::/64 any (2 matches) sequence 10
    permit tcp any any eq 22 (6 matches) sequence 20
IPv6 access list RESTRICTED-LAN
    permit tcp 2001:DB8:ACAD:B::/64 host 2001:DB8:ACAD:A::A eq telnet sequence 15
    deny tcp any 2001:DB8:ACAD:A::/64 eq telnet sequence 20
    permit ipv6 any any (124 matches) sequence 30
    permit tcp any host 2001:DB8:ACAD:A::3 eq www sequence 40
```

Note: The **do** command can be used to execute any privileged EXEC command while in global configuration mode or a submode.

Step 6: Delete an ACL statement.

Use the **no** command to delete the permit statement that you just added.

```
R1(config-ipv6-acl)# no permit tcp any host 2001:DB8:ACAD:A::3 eq www
```

Step 7: Use the do show access-list RESTRICTED-LAN command to view the ACL.

```
R1(config-ipv6-acl)# do show access-list RESTRICTED-LAN
IPv6 access list RESTRICTED-LAN
    permit tcp 2001:DB8:ACAD:B::/64 host 2001:DB8:ACAD:A::A eq telnet sequence 15
    deny tcp any 2001:DB8:ACAD:A::/64 eq telnet sequence 20
    permit ipv6 any any (214 matches) sequence 30
```

Step 8:   Re-apply the RESTRICTED-LAN ACL to the interface G0/1.

```
R1(config-ipv6-acl)# int g0/1

R1(config-if)# ipv6 traffic-filter RESTRICTED-LAN out

R1(config-if)# end
```

Step 9:   Test ACL changes.

Telnet to S1 from PC-B. Troubleshoot if necessary.

Reflection

1.   What is causing the match count on the RESTRICTED-LAN **permit ipv6 any any** statement to continue to increase?

_____

_____

_____

_____

2.   What command would you use to reset the counters for the ACL on the VTY lines?

_____

Router Interface Summary Table

Router Interface Summary				
Router Model	**Ethernet Interface #1**	**Ethernet Interface #2**	**Serial Interface #1**	**Serial Interface #2**
1800	Fast Ethernet 0/0 (F0/0)	Fast Ethernet 0/1 (F0/1)	Serial 0/0/0 (S0/0/0)	Serial 0/0/1 (S0/0/1)
1900	Gigabit Ethernet 0/0 (G0/0)	Gigabit Ethernet 0/1 (G0/1)	Serial 0/0/0 (S0/0/0)	Serial 0/0/1 (S0/0/1)
2801	Fast Ethernet 0/0 (F0/0)	Fast Ethernet 0/1 (F0/1)	Serial 0/1/0 (S0/1/0)	Serial 0/1/1 (S0/1/1)
2811	Fast Ethernet 0/0 (F0/0)	Fast Ethernet 0/1 (F0/1)	Serial 0/0/0 (S0/0/0)	Serial 0/0/1 (S0/0/1)
2900	Gigabit Ethernet 0/0 (G0/0)	Gigabit Ethernet 0/1 (G0/1)	Serial 0/0/0 (S0/0/0)	Serial 0/0/1 (S0/0/1)
Note: To find out how the router is configured, look at the interfaces to identify the type of router and how many interfaces the router has. There is no way to effectively list all the combinations of configurations for each router class. This table includes identifiers for the possible combinations of Ethernet and Serial interfaces in the device. The table does not include any other type of interface, even though a specific router may contain one. An example of this might be an ISDN BRI interface. The string in parenthesis is the legal abbreviation that can be used in Cisco IOS commands to represent the interface.				

9.6.1.1 Class Activity – FTP Denied

Objective

Implement packet filtering using extended IPv4 ACLs according to networking requirements (to include named and numbered ACLs).

Scenario

It was recently reported that viruses are on the rise within your small- to medium-sized business network. Your network administrator has been tracking network performance and has determined that one particular host is constantly downloading files from a remote FTP server. This host just may be the virus source perpetuating throughout the network!

Use Packet Tracer to complete this activity. Write a <u>named</u> ACL to deny the host access to the FTP server. Apply the ACL to the most effective interface on the router.

To complete the physical topology, you must use:

- One PC host station
- Two switches
- One Cisco 1941 series Integrated Services Router
- One server

Using the Packet Tracer text tool, record the ACL you prepared. Validate that the ACL works to deny access to the FTP server by trying to access the FTP server's address. Observe what happens while in simulation mode.

Save your file and be prepared to share it with another student, or with the entire class.

Reflection

1. What was the most difficult part of completing this modeling activity?

    _____

2. How often do you think network administrators need to change their ACLs on their networks?

    _____

3. Why would you consider using a named extended ACL instead of a regular extended ACL?

    _____

Chapter 10 — DHCP

10.0.1.2 Class Activity – Own or Lease?

Objective

Configure DHCP for IPv4 on a LAN switch.

Scenario

This chapter presents the concept of using the DHCP process in a small- to medium-sized business network. This modeling activity describes how very basic wireless ISR devices work using the DHCP process.

Visit http://ui.linksys.com/WRT54GL/4.30.0/Setup.htm, which is a web-based simulator that helps you learn to configure DHCP using a Linksys wireless 54GL router. To the right of the simulator (in the blue description column), you can click **More** to read information about configuring DHCP settings on this particular integrated services router (ISR) simulator.

Practice configuring the ISR's:

- Hostname
- Local IP address with subnet mask
- DHCP (enable and disable)
- Starting IP address
- Maximum number of users to receive an IP DHCP address
- Lease time
- Time zone (use yours or a favorite as an alternative)

When you have completed configuring the settings as listed for this assignment, take a screen shot of your settings by using the **PrtScr** key command. Copy and place your screen shot into a word processing document. Save it and be prepared to discuss your configuration choices with the class.

Required Resources

Internet connectivity

Reflection

1. Why would any network administrator need to save a bank of IP addresses for DHCP **not** to use?

_____

2. You are designing your small- to medium-sized network and you have a choice as to whether to buy a small, generic ISR for DHCP purposes, or use a DHCP full server. Before you read this chapter, how would you make your decision?

_____

10.1.2.4 Lab - Configuring Basic DHCPv4 on a Router

Topology

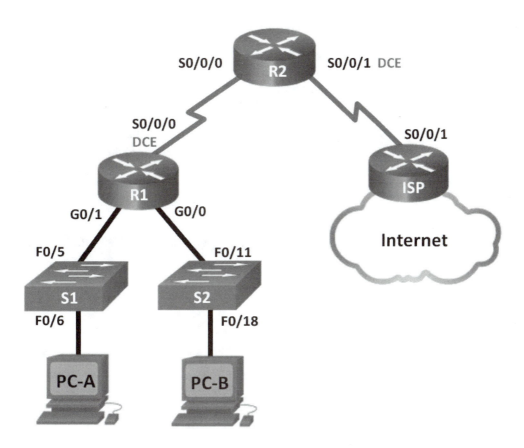

Addressing Table

Device	Interface	IP Address	Subnet Mask	Default Gateway
R1	G0/0	192.168.0.1	255.255.255.0	N/A
	G0/1	192.168.1.1	255.255.255.0	N/A
	S0/0/0 (DCE)	192.168.2.253	255.255.255.252	N/A
R2	S0/0/0	192.168.2.254	255.255.255.252	N/A
	S0/0/1 (DCE)	209.165.200.226	255.255.255.224	N/A
ISP	S0/0/1	209.165.200.225	255.255.255.224	N/A
PC-A	NIC	DHCP	DHCP	DHCP
PC-B	NIC	DHCP	DHCP	DHCP

Objectives

Part 1: Build the Network and Configure Basic Device Settings

Part 2: Configure a DHCPv4 Server and a DHCP Relay Agent

Background / Scenario

The Dynamic Host Configuration Protocol (DHCP) is a network protocol that lets network administrators manage and automate the assignment of IP addresses. Without DHCP, the administrator must manually assign and configure IP addresses, preferred DNS servers, and default gateways. As the network grows in size, this becomes an administrative problem when devices are moved from one internal network to another.

In this scenario, the company has grown in size, and the network administrators can no longer assign IP addresses to devices manually. Your job is to configure the R2 router to assign IPv4 addresses on two different subnets connected to router R1.

Note: This lab provides minimal assistance with the actual commands necessary to configure DHCP. However, the required commands are provided in Appendix A. Test your knowledge by trying to configure the devices without referring to the appendix.

Note: The routers used with CCNA hands-on labs are Cisco 1941 Integrated Services Routers (ISRs) with Cisco IOS Release 15.2(4)M3 (universalk9 image). The switches used are Cisco Catalyst 2960s with Cisco IOS Release 15.0(2) (lanbasek9 image). Other routers, switches and Cisco IOS versions can be used. Depending on the model and Cisco IOS version, the commands available and output produced might vary from what is shown in the labs. Refer to the Router Interface Summary Table at the end of this lab for the correct interface identifiers.

Note: Make sure that the routers and switches have been erased and have no startup configurations. If you are unsure, contact your instructor.

Required Resources

- 3 Routers (Cisco 1941 with Cisco IOS Release 15.2(4)M3 universal image or comparable)
- 2 Switches (Cisco 2960 with Cisco IOS Release 15.0(2) lanbasek9 image or comparable)
- 2 PCs (Windows 7, Vista, or XP with terminal emulation program, such as Tera Term)
- Console cables to configure the Cisco IOS devices via the console ports
- Ethernet and serial cables as shown in the topology

Part 1: Build the Network and Configure Basic Device Settings

In Part 1, you will set up the network topology and configure the routers and switches with basic settings, such as passwords and IP addresses. You will also configure the IP settings for the PCs in the topology.

Step 1: Cable the network as shown in the topology.

Step 2: Initialize and reload the routers and switches.

Step 3: Configure basic settings for each router.

a. Disable DNS lookup

b. Configure the device name as shown in the topology.

c. Assign **class** as the encrypted privileged EXEC mode password.

d. Assign **cisco** as the console and vty passwords.

e. Configure **logging synchronous** to prevent console messages from interrupting command entry.

f. Configure the IP addresses for all the router interfaces according to the Addressing Table.

g. Configure the serial DCE interface on R1 and R2 with a clock rate of 128000.

h. Configure EIGRP for R1.

```
R1(config)# router eigrp 1
R1(config-router)# network 192.168.0.0 0.0.0.255
R1(config-router)# network 192.168.1.0 0.0.0.255
R1(config-router)# network 192.168.2.252 0.0.0.3
R1(config-router)# no auto-summary
```

i. Configure EIGRP and a default route to the ISP on R2.

```
R2(config)# router eigrp 1
R2(config-router)# network 192.168.2.252 0.0.0.3
R2(config-router)# redistribute static
R2(config-router)# exit
R2(config)# ip route 0.0.0.0 0.0.0.0 209.165.200.225
```

j. Configure a summary static route on ISP to reach the networks on the R1 and R2 routers.

```
ISP(config)# ip route 192.168.0.0 255.255.252.0 209.165.200.226
```

k. Copy the running configuration to the startup configuration.

Step 4:   **Verify network connectivity between the routers.**

If any pings between routers fail, correct the errors before proceeding to the next step. Use **show ip route** and **show ip interface brief** to locate possible issues.

Step 5:   **Verify the host PCs are configured for DHCP.**

Part 2:   **Configure a DHCPv4 Server and a DHCP Relay Agent**

To automatically assign address information on the network, you will configure R2 as a DHCPv4 server and R1 as a DHCP relay agent.

Step 1:   **Configure DHCPv4 server settings on router R2.**

On R2, you will configure a DHCP address pool for each of the R1 LANs. Use the pool name **R1G0** for the G0/0 LAN and **R1G1** for the G0/1 LAN. You will also configure the addresses to be excluded from the address pools. Best practice dictates that excluded addresses be configured first, to guarantee that they are not accidentally leased to other devices.

Exclude the first 9 addresses in each R1 LAN starting with .1. All other addresses should be available in the DHCP address pool. Make sure that each DHCP address pool includes a default gateway, the domain **ccna-lab.com**, a DNS server (209.165.200.225), and a lease time of 2 days.

On the lines below, write the commands necessary for configuring DHCP services on router R2, including the DHCP-excluded addresses and the DHCP address pools.

Note: The required commands for Part 2 are provided in Appendix A. Test your knowledge by trying to configure DHCP on R1 and R2 without referring to the appendix.

_____

_____

_____

_____

_____

_____

_____

_____

_____

_____

_____

_____

_____

_____

_____

_____

On PC-A or PC-B, open a command prompt and enter the **ipconfig /all** command. Did either of the host PCs receive an IP address from the DHCP server? Why?

_____

Step 2:   **Configure R1 as a DHCP relay agent.**

Configure IP helper addresses on R1 to forward all DHCP requests to the R2 DHCP server.

On the lines below, write the commands necessary to configure R1 as a DHCP relay agent for the R1 LANs.

_____

_____

_____

_____

Step 3:   **Record IP settings for PC-A and PC-B.**

On PC-A and PC-B, issue the **ipconfig /all** command to verify that the PCs have received IP address information from the DHCP server on R2. Record the IP and MAC address for each PC.

_____

Based on the DHCP pool that was configured on R2, what are the first available IP addresses that PC-A and PC-B can lease?

_____

Step 4:   Verify DHCP services and address leases on R2.

a.   On R2, enter the **show ip dhcp binding** command to view DHCP address leases.

Along with the IP addresses that were leased, what other piece of useful client identification information is in the output?

_____

b.   On R2, enter the **show ip dhcp server statistics** command to view the DHCP pool statistics and message activity.

How many types of DHCP messages are listed in the output?

_____

c.   On R2, enter the **show ip dhcp pool** command to view the DHCP pool settings.

In the output of the **show ip dhcp pool** command, what does the Current index refer to?

_____

d.   On R2, enter the **show run | section dhcp** command to view the DHCP configuration in the running configuration.

e. On R2, enter the **show run interface** command for interfaces G0/0 and G0/1 to view the DHCP relay configuration in the running configuration.

Reflection

What do you think is the benefit of using DHCP relay agents instead of multiple routers acting as DHCP servers?

_____

_____

_____

_____

Router Interface Summary Table

Router Interface Summary				
Router Model	**Ethernet Interface #1**	**Ethernet Interface #2**	**Serial Interface #1**	**Serial Interface #2**
1800	Fast Ethernet 0/0 (F0/0)	Fast Ethernet 0/1 (F0/1)	Serial 0/0/0 (S0/0/0)	Serial 0/0/1 (S0/0/1)
1900	Gigabit Ethernet 0/0 (G0/0)	Gigabit Ethernet 0/1 (G0/1)	Serial 0/0/0 (S0/0/0)	Serial 0/0/1 (S0/0/1)
2801	Fast Ethernet 0/0 (F0/0)	Fast Ethernet 0/1 (F0/1)	Serial 0/1/0 (S0/1/0)	Serial 0/1/1 (S0/1/1)
2811	Fast Ethernet 0/0 (F0/0)	Fast Ethernet 0/1 (F0/1)	Serial 0/0/0 (S0/0/0)	Serial 0/0/1 (S0/0/1)
2900	Gigabit Ethernet 0/0 (G0/0)	Gigabit Ethernet 0/1 (G0/1)	Serial 0/0/0 (S0/0/0)	Serial 0/0/1 (S0/0/1)

Note: To find out how the router is configured, look at the interfaces to identify the type of router and how many interfaces the router has. There is no way to effectively list all the combinations of configurations for each router class. This table includes identifiers for the possible combinations of Ethernet and Serial interfaces in the device. The table does not include any other type of interface, even though a specific router may contain one. An example of this might be an ISDN BRI interface. The string in parenthesis is the legal abbreviation that can be used in Cisco IOS commands to represent the interface.

Appendix A – DHCP Configuration Commands

Router R1

```
R1(config)# interface g0/0
R1(config-if)# ip helper-address 192.168.2.254
R1(config-if)# exit
R1(config-if)# interface g0/1
R1(config-if)# ip helper-address 192.168.2.254
```

Router R2

```
R2(config)# ip dhcp excluded-address 192.168.0.1 192.168.0.9
R2(config)# ip dhcp excluded-address 192.168.1.1 192.168.1.9
R2(config)# ip dhcp pool R1G1
R2(dhcp-config)# network 192.168.1.0 255.255.255.0
R2(dhcp-config)# default-router 192.168.1.1
R2(dhcp-config)# dns-server 209.165.200.225
R2(dhcp-config)# domain-name ccna-lab.com
R2(dhcp-config)# lease 2
R2(dhcp-config)# exit
R2(config)# ip dhcp pool R1G0
R2(dhcp-config)# network 192.168.0.0 255.255.255.0
R2(dhcp-config)# default-router 192.168.0.1
R2(dhcp-config)# dns-server 209.165.200.225
R2(dhcp-config)# domain-name ccna-lab.com
R2(dhcp-config)# lease 2
```

10.1.2.5 Lab – Configuring Basic DHCPv4 on a Switch

Topology

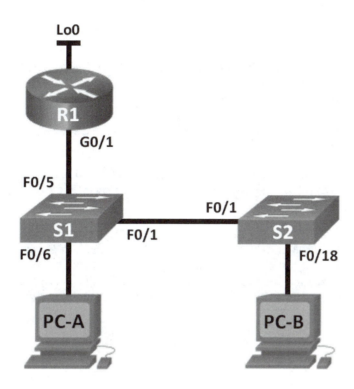

Addressing Table

Device	Interface	IP Address	Subnet Mask
R1	G0/1	192.168.1.10	255.255.255.0
	Lo0	209.165.200.225	255.255.255.224
S1	VLAN 1	192.168.1.1	255.255.255.0
	VLAN 2	192.168.2.1	255.255.255.0

Objectives

Part 1: Build the Network and Configure Basic Device Settings

Part 2: Change the SDM Preference

- Set the SDM preference to lanbase-routing on S1.

Part 3: Configure DHCPv4

- Configure DHCPv4 for VLAN 1.
- Verify DHCPv4 and connectivity.

Part 4: Configure DHCP for Multiple VLANs

- Assign ports to VLAN 2.
- Configure DHCPv4 for VLAN 2.
- Verify DHCPv4 and connectivity.

Part 5: Enable IP Routing

- Enable IP routing on the switch.

- Create static routes.

Background / Scenario

A Cisco 2960 switch can function as a DHCPv4 server. The Cisco DHCPv4 server assigns and manages IPv4 addresses from identified address pools that are associated with specific VLANs and switch virtual interfaces (SVIs). The Cisco 2960 switch can also function as a Layer 3 device and route between VLANs and a limited number of static routes. In this lab, you will configure DHCPv4 for both single and multiple VLANs on a Cisco 2960 switch, enable routing on the switch to allow for communication between VLANs, and add static routes to allow for communication between all hosts.

Note: This lab provides minimal assistance with the actual commands necessary to configure DHCP. However, the required commands are provided in Appendix A. Test your knowledge by trying to configure the devices without referring to the appendix.

Note: The routers used with CCNA hands-on labs are Cisco 1941 Integrated Services Routers (ISRs) with Cisco IOS Release 15.2(4)M3 (universalk9 image). The switches used are Cisco Catalyst 2960s with Cisco IOS Release 15.0(2) (lanbasek9 image). Other routers, switches and Cisco IOS versions can be used. Depending on the model and Cisco IOS version, the commands available and output produced might vary from what is shown in the labs. Refer to the Router Interface Summary Table at the end of this lab for the correct interface identifiers.

Note: Make sure that the router and switches have been erased and have no startup configurations. If you are unsure, contact your instructor.

Required Resources

- 1 Router (Cisco 1941 with Cisco IOS Release 15.2(4)M3 universal image or comparable)

- 2 Switches (Cisco 2960 with Cisco IOS Release 15.0(2) lanbasek9 image or comparable)

- 2 PCs (Windows 7, Vista, or XP with terminal emulation program, such as Tera Term)

- Console cables to configure the Cisco IOS devices via the console ports

- Ethernet cables as shown in the topology

Part 1: Build the Network and Configure Basic Device Settings

Step 1: **Cable the network as shown in the topology.**

Step 2: **Initialize and reload the router and switches.**

Step 3: **Configure basic setting on devices.**

a. Assign device names as shown in the topology.

b. Disable DNS lookup.

c. Assign **class** as the enable password and assign **cisco** as the console and vty passwords.

d. Configure the IP addresses on R1 G0/1 and Lo0 interfaces, according to the Addressing Table.

e. Configure the IP addresses on S1 VLAN 1 and VLAN 2 interfaces, according to the Addressing Table.

f. Save the running configuration to the startup configuration file.

Part 2: Change the SDM Preference

The Cisco Switch Database Manager (SDM) provides multiple templates for the Cisco 2960 switch. The templates can be enabled to support specific roles depending on how the switch is used in the network. In this lab, the sdm lanbase-routing template is enabled to allow the switch to route between VLANs and to support static routing.

Step 1: **Display the SDM preference on S1.**

On S1, issue the **show sdm prefer** command in privileged EXEC mode. If the template has not been changed from the factory default, it should still be the **default** template. The **default** template does not support static routing. If IPv6 addressing has been enabled, the template will be **dual-ipv4-and-ipv6 default**.

```
S1# show sdm prefer
The current template is "default" template.
The selected template optimizes the resources in
the switch to support this level of features for
0 routed interfaces and 255 VLANs.

    number of unicast mac addresses:           8K
    number of IPv4 IGMP groups:                0.25K
    number of IPv4/MAC qos aces:               0.125k
    number of IPv4/MAC security aces:          0.375k
```

What is the current template?

Step 2: **Change the SDM Preference on S1.**

a. Set the SDM preference to **lanbase-routing**. (If lanbase-routing is the current template, please proceed to Part 3.) From global configuration mode, issue the **sdm prefer lanbase-routing** command.

```
S1(config)# sdm prefer lanbase-routing
Changes to the running SDM preferences have been stored, but cannot take effect
until the next reload.
Use 'show sdm prefer' to see what SDM preference is currently active.
```

Which template will be available after reload? _____

b. The switch must be reloaded for the template to be enabled.

```
S1# reload
```

```
System configuration has been modified. Save? [yes/no]: no
Proceed with reload? [confirm]
```

Note: The new template will be used after reboot even if the running configuration has not been saved. To save the running configuration, answer **yes** to save the modified system configuration.

Step 3: **Verify that lanbase-routing template is loaded.**

Issue the **show sdm prefer** command to verify that the lanbase-routing template has been loaded on S1.

```
S1# show sdm prefer
 The current template is "lanbase-routing" template.
 The selected template optimizes the resources in
 the switch to support this level of features for
 0 routed interfaces and 255 VLANs.

  number of unicast mac addresses:                 4K
  number of IPv4 IGMP groups + multicast routes:   0.25K
  number of IPv4 unicast routes:                   0.75K
    number of directly-connected IPv4 hosts:       0.75K
```

```
number of indirect IPv4 routes:              16

number of IPv6 multicast groups:             0.375k

number of directly-connected IPv6 addresses: 0.75K

   number of indirect IPv6 unicast routes:   16

number of IPv4 policy based routing aces:    0

number of IPv4/MAC qos aces:                 0.125k

number of IPv4/MAC security aces:            0.375k

number of IPv6 policy based routing aces:    0

number of IPv6 qos aces:                     0.375k

number of IPv6 security aces:                127
```

Part 3: Configure DHCPv4

In Part 3, you will configure DHCPv4 for VLAN 1, check IP settings on host computers to validate DHCP functionality, and verify connectivity for all devices in VLAN 1.

Step 1: Configure DHCP for VLAN 1.

a. Exclude the first 10 valid host addresses from network 192.168.1.0/24. Write the command you used in the space provided.

_____

b. Create a DHCP pool named **DHCP1**. Write the command you used in the space provided.

_____

c. Assign the network 192.168.1.0/24 for available addresses. Write the command you used in the space provided.

_____

d. Assign the default gateway as 192.168.1.1. Write the command you used in the space provided.

_____

e. Assign the DNS server as 192.168.1.9. Write the command you used in the space provided.

_____

f. Assign a lease time of 3 days. Write the command you used in the space provided.

_____

g. Save the running configuration to the startup configuration file.

Step 2:   **Verify DHCP and connectivity.**

a.   On PC-A and PC-B, open the command prompt and issue the **ipconfig** command. If IP information is not present, or if it is incomplete, issue the **ipconfig /release** command, followed by the **ipconfig /renew** command.

For PC-A, list the following:

IP Address: _____

Subnet Mask: _____

Default Gateway: _____

For PC-B, list the following:

IP Address: _____

Subnet Mask: _____

Default Gateway: _____

b.   Test connectivity by pinging from PC-A to the default gateway, PC-B, and R1.

From PC-A, is it possible to ping the VLAN 1 default gateway? _____

From PC-A, is it possible to ping PC-B? _____

From PC-A, is it possible to ping R1 G0/1? _____

If the answer is no to any of these questions, troubleshoot the configurations and correct the error.

Part 4:   Configure DHCPv4 for Multiple VLANs

In Part 4, you will assign PC-A to a port accessing VLAN 2, configure DHCPv4 for VLAN 2, renew the IP configuration of PC-A to validate DHCPv4, and verify connectivity within the VLAN.

Step 1:   **Assign a port to VLAN 2.**

Place port F0/6 into VLAN 2. Write the command you used in the space provided.

_____

_____

Step 2:   **Configure DHCPv4 for VLAN 2**

a.   Exclude the first 10 valid host addresses from network 192.168.2.0. Write the command you used in the space provided.

_____

b.   Create a DHCP pool named **DHCP2**. Write the command you used in the space provided.

_____

c. Assign the network 192.168.2.0/24 for available addresses. Write the command you used in the space provided.

_____

d. Assign the default gateway as 192.168.2.1. Write the command you used in the space provided.

_____

e. Assign the DNS server as 192.168.2.9. Write the command you used in the space provided.

_____

f. Assign a lease time of 3 days. Write the command you used in the space provided.

_____

g. Save the running configuration to the startup configuration file.

Step 3: **Verify DHCPv4 and connectivity.**

a. On PC-A, open the command prompt and issue the **ipconfig /release** command, followed by **ipconfig / renew** command.

For PC-A, list the following:

IP Address: _____

Subnet Mask: _____

Default Gateway: _____

b. Test connectivity by pinging from PC-A to the VLAN 2 default gateway and PC-B.

From PC-A, is it possible to ping the default gateway? _____

From PC-A, is it possible to ping PC-B? _____

Were these pings successful? Why?

_____

_____

c. Issue the **show ip route** command on S1.

What was the result of this command?

_____

_____

Part 5: Enable IP Routing

In Part 5, you will enable IP routing on the switch, which will allow for inter-VLAN communication. For all networks to communicate, static routes on S1 and R1 must be implemented.

Step 1:   Enable IP routing on S1.

a. From global configuration mode, use the **ip routing** command to enable routing on S1.

```
S1(config)# ip routing
```

b. Verify inter-VLAN connectivity.

From PC-A, is it possible to ping PC-B? _____

What function is the switch performing?

_____

c. View the routing table information for S1.

What route information is contained in the output of this command?

_____

_____

d. View the routing table information for R1.

What route information is contained in the output of this command?

_____

_____

e. From PC-A, is it possible to ping R1? _____

From PC-A, is it possible to ping Lo0? _____

Consider the routing table of the two devices, what must be added to communicate between all networks?

_____

Step 2:   **Assign static routes.**

Enabling IP routing allows the switch to route between VLANs assigned on the switch. For all VLANs to com-municate with the router, static routes must be added to the routing table of both the switch and the router.

a. On S1, create a default static route to R1. Write the command you used in the space provided.

_____

b. On R1, create a static route to VLAN 2. Write the command you used in the space provided.

_____

c. View the routing table information for S1.

How is the default static route represented?

_____

d. View the routing table information for R1.

How is the static route represented?

_____

e. From PC-A, is it possible to ping R1? _____

From PC-A, is it possible to ping Lo0? _____

Reflection

1. In configuring DHCPv4, why would you exclude the static addresses prior to setting up the DHCPv4 pool?

_____

_____

2. If multiple DHCPv4 pools are present, how does the switch assign the IP information to hosts?

_____

3. Besides switching, what functions can the Cisco 2960 switch perform?

_____

Router Interface Summary Table

Router Interface Summary				
Router Model	**Ethernet Interface #1**	**Ethernet Interface #2**	**Serial Interface #1**	**Serial Interface #2**
1800	Fast Ethernet 0/0 (F0/0)	Fast Ethernet 0/1 (F0/1)	Serial 0/0/0 (S0/0/0)	Serial 0/0/1 (S0/0/1)
1900	Gigabit Ethernet 0/0 (G0/0)	Gigabit Ethernet 0/1 (G0/1)	Serial 0/0/0 (S0/0/0)	Serial 0/0/1 (S0/0/1)
2801	Fast Ethernet 0/0 (F0/0)	Fast Ethernet 0/1 (F0/1)	Serial 0/1/0 (S0/1/0)	Serial 0/1/1 (S0/1/1)
2811	Fast Ethernet 0/0 (F0/0)	Fast Ethernet 0/1 (F0/1)	Serial 0/0/0 (S0/0/0)	Serial 0/0/1 (S0/0/1)
2900	Gigabit Ethernet 0/0 (G0/0)	Gigabit Ethernet 0/1 (G0/1)	Serial 0/0/0 (S0/0/0)	Serial 0/0/1 (S0/0/1)

Note: To find out how the router is configured, look at the interfaces to identify the type of router and how many interfaces the router has. There is no way to effectively list all the combinations of configurations for each router class. This table includes identifiers for the possible combinations of Ethernet and Serial interfaces in the device. The table does not include any other type of interface, even though a specific router may contain one. An example of this might be an ISDN BRI interface. The string in parenthesis is the legal abbreviation that can be used in Cisco IOS commands to represent the interface.

Appendix A: Configuration Commands

Configure DHCPv4

```
S1(config)# ip dhcp excluded-address 192.168.1.1 192.168.1.10

S1(config)# ip dhcp pool DHCP1

S1(dhcp-config)# network 192.168.1.0 255.255.255.0

S1(dhcp-config)# default-router 192.168.1.1

S1(dhcp-config)# dns-server 192.168.1.9

S1(dhcp-config)# lease 3
```

Configure DHCPv4 for Multiple VLANs

```
S1(config)# interface f0/6

S1(config-if)# switchport access vlan 2

S1(config)# ip dhcp excluded-address 192.168.2.1 192.168.2.10

S1(config)# ip dhcp pool DHCP2

S1(dhcp-config)# network 192.168.2.0 255.255.255.0

S1(dhcp-config)# default-router 192.168.2.1

S1(dhcp-config)# dns-server 192.168.2.9

S1(dhcp-config)# lease 3
```

Enable IP Routing

```
S1(config)# ip routing

S1(config)# ip route 0.0.0.0 0.0.0.0 192.168.1.10

R1(config)# ip route 192.168.2.0 255.255.255.0 g0/1
```

10.1.4.4 Lab - Troubleshooting DHCPv4

Topology

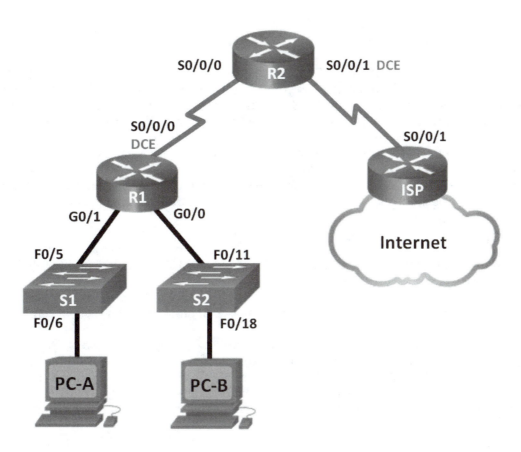

Addressing Table

Device	Interface	IP Address	Subnet Mask	Default Gateway
R1	G0/0	192.168.0.1	255.255.255.128	N/A
	G0/1	192.168.1.1	255.255.255.0	N/A
	S0/0/0 (DCE)	192.168.0.253	255.255.255.252	N/A
R2	S0/0/0	192.168.0.254	255.255.255.252	N/A
	S0/0/1 (DCE)	209.165.200.226	255.255.255.252	N/A
ISP	S0/0/1	209.165.200.225	255.255.255.252	N/A
S1	VLAN 1	192.168.1.2	255.255.255.0	192.168.1.1
S2	VLAN 1	192.168.0.2	255.255.255.128	192.168.0.1
PC-A	NIC	DHCP	DHCP	DHCP
PC-B	NIC	DHCP	DHCP	DHCP

Objectives

Part 1: Build the Network and Configure Basic Device Settings

Part 2: Troubleshoot DHCPv4 Issues

Background / Scenario

The Dynamic Host Configuration Protocol (DHCP) is a network protocol that lets the network administrators manage and automate the assignment of IP addresses. Without DHCP, the administrator must manually assign and configure IP addresses, preferred DNS servers, and the default gateway. As the network grows in size, this becomes an administrative problem when devices are moved from one internal network to another.

In this scenario, the company has grown in size, and the network administrators can no longer assign IP addresses to devices manually. The R2 router has been configured as a DHCP server to assign IP addresses to the host devices on router R1 LANs. Several errors in the configuration have resulted in connectivity issues. You are asked to troubleshoot and correct the configuration errors and document your work.

Ensure that the network supports the following:

1) The router R2 should function as the DHCP server for the 192.168.0.0/25 and 192.168.1.0/24 networks connected to R1.

2) All PCs connected to S1 and S2 should receive an IP address in the correct network via DHCP.

Note: The routers used with CCNA hands-on labs are Cisco 1941 Integrated Services Routers (ISRs) with Cisco IOS Release 15.2(4)M3 (universalk9 image). The switches used are Cisco Catalyst 2960s with Cisco IOS Release 15.0(2) (lanbasek9 image). Other routers, switches and Cisco IOS versions can be used. Depending on the model and Cisco IOS version, the commands available and output produced might vary from what is shown in the labs. Refer to the Router Interface Summary Table at the end of this lab for the correct interface identifiers.

Note: Make sure that the routers and switches have been erased and have no startup configurations. If you are unsure, contact your instructor.

Required Resources

- 3 Routers (Cisco 1941 with Cisco IOS Release 15.2(4)M3 universal image or comparable)
- 2 Switches (Cisco 2960 with Cisco IOS Release 15.0(2) lanbasek9 image or comparable)
- 2 PCs (Windows 7, Vista, or XP with terminal emulation program, such as Tera Term)
- Console cables to configure the Cisco IOS devices via the console ports
- Ethernet and serial cables as shown in the topology

Part 1:   Build the Network and Configure Basic Device Settings

In Part 1, you will set up the network topology and configure the routers and switches with basic settings, such as passwords and IP addresses. You will also configure the IP settings for the PCs in the topology.

Step 1:   Cable the network as shown in the topology.

Step 2:   Initialize and reload the routers and switches.

Step 3:   Configure basic settings for each router.

a.   Disable DNS lookup.

b.   Configure device name as shown in the topology.

c.   Assign **class** as the privileged EXEC password.

d.   Assign **cisco** as the console and vty passwords.

e.   Configure **logging synchronous** to prevent console messages from interrupting command entry.

f.   Configure the IP addresses for all the router interfaces.

g.   Set clock rate to **128000** for all DCE router interfaces.

h.   Configure EIGRP for R1.

```
R1(config)# router eigrp 1
R1(config-router)# network 192.168.0.0 0.0.0.127
R1(config-router)# network 192.168.0.252 0.0.0.3
R1(config-router)# network 192.168.1.0
R1(config-router)# no auto-summary
```

i.   Configure EIGRP and a static default route on R2.

```
R2(config)# router eigrp 1
R2(config-router)# network 192.168.0.252 0.0.0.3
R2(config-router)# redistribute static
R2(config-router)# exit
R2(config)# ip route 0.0.0.0 0.0.0.0 209.165.200.225
```

j.   Configure a summary static route on ISP to the networks on R1 and R2 routers.

```
ISP(config)# ip route 192.168.0.0 255.255.254.0 209.165.200.226
```

Step 4:   Verify network connectivity between the routers.

If any pings between the routers fail, correct the errors before proceeding to the next step. Use **show ip route** and **show ip interface brief** to locate possible issues.

Step 5:   Configure basic settings for each switch.

a.   Disable DNS lookup.

b.   Configure device name as shown in the topology.

c.   Configure the IP address for the VLAN 1 interface and the default gateway for each switch.

  d. Assign **class** as the privileged EXEC mode password.

  e. Assign **cisco** as the console and vty passwords.

  f. Configure **logging synchronous** for the console line.

Step 6: **Verify the hosts are configured for DHCP.**

Step 7: **Load the initial DHCP configuration for R1 and R2.**

Router R1

```
interface GigabitEthernet0/1
 ip helper-address 192.168.0.253
```

Router R2

```
ip dhcp excluded-address 192.168.11.1 192.168.11.9

ip dhcp excluded-address 192.168.0.1 192.168.0.9
ip dhcp pool R1G1
 network 192.168.1.0 255.255.255.0
 default-router 192.168.1.1
ip dhcp pool R1G0
 network 192.168.0.0 255.255.255.128
 default-router 192.168.11.1
```

Part 2: **Troubleshoot DHCPv4 Issues**

After configuring routers R1 and R2 with DHCPv4 settings, several errors in the DHCP configurations were introduced and resulted in connectivity issues. R2 is configured as a DHCP server. For both pools of DHCP addresses, the first nine addresses are reserved for the routers and switches. R1 relays the DHCP information to all the R1 LANs. Currently, PC-A and PC-B have no access to the network. Use the **show** and **debug** commands to determine and correct the network connectivity issues.

Step 1: **Record IP settings for PC-A and PC-B.**

  a. For PC-A and PC-B, at the command prompt, enter **ipconfig /all** to display the IP and MAC addresses.

  b. Record the IP and MAC addresses in the table below. The MAC address can be used to determine which PC is involved in the debug message.

	IP Address/Subnet Mask	MAC Address
PC-A		
PC-B		

Step 2:   Troubleshoot DHCP issues for the 192.168.1.0/24 network on router R1.

Router R1 is a DHCP relay agent for all the R1 LANs. In this step, only the DHCP process for the 192.168.1.0/24 network will be examined. The first nine addresses are reserved for other network devices, such as routers, switches, and servers.

a. Use a DHCP **debug** command to observe the DHCP process on R2 router.

```
R2# debug ip dhcp server events
```

b. On R1, display the running configuration for the G0/1 interface.

```
R1# show run interface g0/1
interface GigabitEthernet0/1
 ip address 192.168.1.1 255.255.255.0
 ip helper-address 192.168.0.253
 duplex auto
 speed auto
```

If there are any DHCP relay issues, record any commands that are necessary to correct the configurations errors.

_____

_____

_____

c. In a command prompt on PC-A, type **ipconfig /renew** to receive an address from the DHCP server. Record the configured IP address, subnet mask, and default gateway for PC-A.

_____

d. Observe the debug messages on R2 router for the DHCP renewal process for PC-A. The DHCP server attempted to assign 192.168.1.1/24 to PC-A. This address is already in use for G0/1 interface on R1. The same issue occurs with IP address 192.168.1.2/24 because this address has been assigned to S1 in the initial configuration. Therefore, an IP address of 192.168.1.3/24 has been assigned to PC-A. The DHCP assignment conflict indicates there may be an issue with the excluded-address statement on the DHCP server configuration on R2.

```
*Mar  5 06:32:16.939: DHCPD: Sending notification of DISCOVER:

*Mar  5 06:32:16.939:   DHCPD: htype 1 chaddr 0050.56be.768c

*Mar  5 06:32:16.939:   DHCPD: circuit id 00000000

*Mar  5 06:32:16.939: DHCPD: Seeing if there is an internally specified pool class:

*Mar  5 06:32:16.939:   DHCPD: htype 1 chaddr 0050.56be.768c

*Mar  5 06:32:16.939:   DHCPD: circuit id 00000000

*Mar  5 06:32:16.943: DHCPD: Allocated binding 2944C764

*Mar  5 06:32:16.943: DHCPD: Adding binding to radix tree (192.168.1.1)

*Mar  5 06:32:16.943: DHCPD: Adding binding to hash tree

*Mar  5 06:32:16.943: DHCPD: assigned IP address 192.168.1.1 to client 0100.5056.
be76.8c.

*Mar  5 06:32:16.951: %DHCPD-4-PING_CONFLICT: DHCP address conflict:  server pinged
192.168.1.1.

*Mar  5 06:32:16.951: DHCPD: returned 192.168.1.1 to address pool R1G1.

*Mar  5 06:32:16.951: DHCPD: Sending notification of DISCOVER:

*Mar  5 06:32:16.951:   DHCPD: htype 1 chaddr 0050.56be.768c

*Mar  5 06:32:16.951:   DHCPD: circuit id 00000000

*Mar  5 06:32:1

R2#6.951: DHCPD: Seeing if there is an internally specified pool class:

*Mar  5 06:32:16.951:   DHCPD: htype 1 chaddr 0050.56be.768c

*Mar  5 06:32:16.951:   DHCPD: circuit id 00000000

*Mar  5 06:32:16.951: DHCPD: Allocated binding 31DC93C8

*Mar  5 06:32:16.951: DHCPD: Adding binding to radix tree (192.168.1.2)

*Mar  5 06:32:16.951: DHCPD: Adding binding to hash tree

*Mar  5 06:32:16.951: DHCPD: assigned IP address 192.168.1.2 to client 0100.5056.
be76.8c.

*Mar  5 06:32:18.383: %DHCPD-4-PING_CONFLICT: DHCP address conflict:  server pinged
192.168.1.2.

*Mar  5 06:32:18.383: DHCPD: returned 192.168.1.2 to address pool R1G1.

*Mar  5 06:32:18.383: DHCPD: Sending notification of DISCOVER:

*Mar  5 06:32:18.383:   DHCPD: htype 1 chaddr 0050.56be.6c89

*Mar  5 06:32:18.383:   DHCPD: circuit id 00000000

*Mar  5 06:32:18.383: DHCPD: Seeing if there is an internally specified pool class:

*Mar  5 06:32:18.383:   DHCPD: htype 1 chaddr 0050.56be.6c89

*Mar  5 06:32:18.383:   DHCPD: circuit id 00000000

*Mar  5 06:32:18.383: DHCPD: Allocated binding 2A40E074

*Mar  5 06:32:18.383: DHCPD: Adding binding to radix tree (192.168.1.3)

*Mar  5 06:32:18.383: DHCPD: Adding binding to hash tree

*Mar  5 06:32:18.383: DHCPD: assigned IP address 192.168.1.3 to client 0100.5056.
be76.8c.

<output omitted>
```

e. Display the DHCP server configuration on R2. The first nine addresses for 192.168.1.0/24 network are not excluded from the DHCP pool.

```
R2# show run | section dhcp
ip dhcp excluded-address 192.168.11.1 192.168.11.9
ip dhcp excluded-address 192.168.0.1 192.168.0.9
ip dhcp pool R1G1
 network 192.168.1.0 255.255.255.0
 default-router 192.168.1.1
ip dhcp pool R1G0
 network 192.168.0.0 255.255.255.128
 default-router 192.168.1.1
```

Record the commands to resolve the issue on R2.

_____

_____

f. At the command prompt on PC-A, type **ipconfig /release** to return the 192.168.1.3 address back to the DHCP pool. The process can be observed in the debug message on R2.

```
*Mar  5 06:49:59.563: DHCPD: Sending notification of TERMINATION:
*Mar  5 06:49:59.563:  DHCPD: address 192.168.1.3 mask 255.255.255.0
*Mar  5 06:49:59.563:  DHCPD: reason flags: RELEASE
*Mar  5 06:49:59.563:   DHCPD: htype 1 chaddr 0050.56be.768c
*Mar  5 06:49:59.563:   DHCPD: lease time remaining (secs) = 85340
*Mar  5 06:49:59.563: DHCPD: returned 192.168.1.3 to address pool R1G1.
```

g. At the command prompt on PC-A, type **ipconfig /renew** to be assigned a new IP address from the DHCP server. Record the assigned IP address and default gateway information.

_____

_____

The process can be observed in the debug message on R2.

```
*Mar  5 06:50:11.863: DHCPD: Sending notification of DISCOVER:
*Mar  5 06:50:11.863:   DHCPD: htype 1 chaddr 0050.56be.768c
*Mar  5 06:50:11.863:   DHCPD: circuit id 00000000
*Mar  5 06:50:11.863: DHCPD: Seeing if there is an internally specified pool class:
*Mar  5 06:50:11.863:   DHCPD: htype 1 chaddr 0050.56be.768c
*Mar  5 06:50:11.863:   DHCPD: circuit id 00000000
*Mar  5 06:50:11.863: DHCPD: requested address 192.168.1.3 has already been assigned.
*Mar  5 06:50:11.863: DHCPD: Allocated binding 3003018C
*Mar  5 06:50:11.863: DHCPD: Adding binding to radix tree (192.168.1.10)
*Mar  5 06:50:11.863: DHCPD: Adding binding to hash tree
*Mar  5 06:50:11.863: DHCPD: assigned IP address 192.168.1.10 to client 0100.5056.
be76.8c.

<output omitted>
```

h. Verify network connectivity.

Can PC-A ping the assigned default gateway? _____

Can PC-A ping the R2 router? _____

Can PC-A ping the ISP router? _____

Step 3: Troubleshoot DHCP issues for 192.168.0.0/25 network on R1.

Router R1 is a DHCP relay agent for all the R1 LANs. In this step, only the DHCP process for the 192.168.0.0/25 network is examined. The first nine addresses are reserved for other network devices.

a. Use a DHCP **debug** command to observe the DHCP process on R2.

```
R2# debug ip dhcp server events
```

b. Display the running configuration for the G0/0 interface on R1 to identify possible DHCP issues.

```
R1# show run interface g0/0
interface GigabitEthernet0/0
 ip address 192.168.0.1 255.255.255.128
 duplex auto
 speed auto
```

Record the issues and any commands that are necessary to correct the configurations errors.

_____

_____

_____

c. From the command prompt on PC-B, type **ipconfig /renew** to receive an address from the DHCP server. Record the configured IP address, subnet mask, and default gateway for PC-B.

_____

d. Observe the debug messages on R2 router for the renewal process for PC-A. The DHCP server assigned 192.168.0.10/25 to PC-B.

```
*Mar  5 07:15:09.663: DHCPD: Sending notification of DISCOVER:
*Mar  5 07:15:09.663:   DHCPD: htype 1 chaddr 0050.56be.f6db
*Mar  5 07:15:09.663:   DHCPD: circuit id 00000000
*Mar  5 07:15:09.663: DHCPD: Seeing if there is an internally specified pool class:
*Mar  5 07:15:09.663:   DHCPD: htype 1 chaddr 0050.56be.f6db
*Mar  5 07:15:09.663:   DHCPD: circuit id 00000000
*Mar  5 07:15:09.707: DHCPD: Sending notification of ASSIGNMENT:
*Mar  5 07:15:09.707:   DHCPD: address 192.168.0.10 mask 255.255.255.128
*Mar  5 07:15:09.707:   DHCPD: htype 1 chaddr 0050.56be.f6db
*Mar  5 07:15:09.707:   DHCPD: lease time remaining (secs) = 86400
```

e. Verify network connectivity.

Can PC-B ping the DHCP assigned default gateway? _____

Can PC-B ping its default gateway (192.168.0.1)? _____

Can PC-B ping the R2 router? _____

Can PC-B ping the ISP router? _____

f. If any issues failed in Step e, record the problems and any commands to resolve the issues.

_____

_____

g. Release and renew the IP configurations on PC-B. Repeat Step e to verify network connectivity.

h. Discontinue the debug process by using the **undebug all** command.

```
R2# undebug all
All possible debugging has been turned off
```

Reflection

What are the benefits of using DHCP?

_____

_____

Router Interface Summary Table

Router Interface Summary				
Router Model	**Ethernet Interface #1**	**Ethernet Interface #2**	**Serial Interface #1**	**Serial Interface #2**
1800	Fast Ethernet 0/0 (F0/0)	Fast Ethernet 0/1 (F0/1)	Serial 0/0/0 (S0/0/0)	Serial 0/0/1 (S0/0/1)
1900	Gigabit Ethernet 0/0 (G0/0)	Gigabit Ethernet 0/1 (G0/1)	Serial 0/0/0 (S0/0/0)	Serial 0/0/1 (S0/0/1)
2801	Fast Ethernet 0/0 (F0/0)	Fast Ethernet 0/1 (F0/1)	Serial 0/1/0 (S0/1/0)	Serial 0/1/1 (S0/1/1)
2811	Fast Ethernet 0/0 (F0/0)	Fast Ethernet 0/1 (F0/1)	Serial 0/0/0 (S0/0/0)	Serial 0/0/1 (S0/0/1)
2900	Gigabit Ethernet 0/0 (G0/0)	Gigabit Ethernet 0/1 (G0/1)	Serial 0/0/0 (S0/0/0)	Serial 0/0/1 (S0/0/1)

Note: To find out how the router is configured, look at the interfaces to identify the type of router and how many interfaces the router has. There is no way to effectively list all the combinations of configurations for each router class. This table includes identifiers for the possible combinations of Ethernet and Serial interfaces in the device. The table does not include any other type of interface, even though a specific router may contain one. An example of this might be an ISDN BRI interface. The string in parenthesis is the legal abbreviation that can be used in Cisco IOS commands to represent the interface.

10.2.3.5 Lab – Configuring Stateless and Stateful DHCPv6

Topology

Addressing Table

Device	Interface	IPv6 Address	Prefix Length	Default Gateway
R1	G0/1	2001:DB8:ACAD:A::1	64	N/A
S1	VLAN 1	Assigned by SLAAC	64	Assigned by SLAAC
PC-A	NIC	Assigned by SLAAC and DHCPv6	64	Assigned by R1

Objectives

Part 1: Build the Network and Configure Basic Device Settings

Part 2: Configure the Network for SLAAC

Part 3: Configure the Network for Stateless DHCPv6

Part 4: Configure the Network for Stateful DHCPv6

Background / Scenario

The dynamic assignment of IPv6 global unicast addresses can be configured in three ways:

- Stateless Address Autoconfiguration (SLAAC) only
- Stateless Dynamic Host Configuration Protocol for IPv6 (DHCPv6)
- Stateful DHCPv6

With SLAAC (pronounced slack), a DHCPv6 server is not needed for hosts to acquire IPv6 addresses. It can be used to receive additional information that the host needs, such as the domain name and the domain name server (DNS) address. When SLAAC is used to assign the IPv6 host addresses and DHCPv6 is used to assign other network parameters, it is called Stateless DHCPv6.

With Stateful DHCPv6, the DHCP server assigns all information, including the host IPv6 address.

Determination of how hosts obtain their dynamic IPv6 addressing information is dependent on flag settings contained within the router advertisement (RA) messages.

In this lab, you will initially configure the network to use SLAAC. After connectivity has been verified, you will configure DHCPv6 settings and change the network to use Stateless DHCPv6. After verification that Stateless DHCPv6 is functioning correctly, you will change the configuration on R1 to use Stateful DHCPv6. Wireshark will be used on PC-A to verify all three dynamic network configurations.

Note: The routers used with CCNA hands-on labs are Cisco 1941 Integrated Services Routers (ISRs) with Cisco IOS Release 15.2(4)M3 (universalk9 image). The switches used are Cisco Catalyst 2960s with Cisco IOS Release 15.0(2) (lanbasek9 image). Other routers, switches and Cisco IOS versions can be used. Depending on the model and Cisco IOS version, the commands available and output produced might vary from what is shown in the labs. Refer to the Router Interface Summary Table at the end of this lab for the correct interface identifiers.

Note: Make sure that the router and switch have been erased and have no startup configurations. If you are unsure, contact your instructor.

Note: The **default bias** template (used by the Switch Database Manager (SDM)) does not provide IPv6 address capabilities. Verify that SDM is using either the **dual-ipv4-and-ipv6** template or the **lanbase-routing** template. The new template will be used after reboot even if the config is not saved.

```
S1# show sdm prefer
```

Follow these steps to assign the **dual-ipv4-and-ipv6** template as the default SDM template:

```
S1# config t
S1(config)# sdm prefer dual-ipv4-and-ipv6 default
S1(config)# end
S1# reload
```

Required Resources

- 1 Router (Cisco 1941 with Cisco IOS Release 15.2(4)M3 universal image or comparable)
- 1 Switch (Cisco 2960 with Cisco IOS Release 15.0(2) lanbasek9 image or comparable)
- 1 PC (Windows 7 or Vista with Wireshark and terminal emulation program, such as Tera Term)
- Console cables to configure the Cisco IOS devices via the console ports
- Ethernet cables as shown in the topology

Note: DHCPv6 client services are disabled on Windows XP. It is recommended to use a Windows 7 host for this lab.

Part 1:   Build the Network and Configure Basic Device Settings

In Part 1, you will set up the network topology and configure basic settings, such as device names, passwords and interface IP addresses.

Step 1:   Cable the network as shown in the topology.

Step 2:   Initialize and reload the router and switch as necessary.

Step 3:   Configure R1.

a. Disable DNS lookup.

b. Configure the device name.

c.   Encrypt plain text passwords.

d.   Create a MOTD banner warning users that unauthorized access is prohibited.

e.   Assign **class** as the encrypted privileged EXEC mode password.

f.   Assign **cisco** as the console and vty password and enable login.

g.   Set console logging to synchronous mode.

h.   Save the running configuration to the startup configuration.

Step 4:   **Configure S1.**

a.   Disable DNS lookup.

b.   Configure the device name.

c.   Encrypt plain text passwords.

d.   Create a MOTD banner warning users that unauthorized access is prohibited.

e.   Assign **class** as the encrypted privileged EXEC mode password.

f.   Assign **cisco** as the console and vty password and enable login.

g.   Set console logging to synchronous mode.

h.   Administratively disable all inactive interfaces.

i.   Save running configuration to the startup configuration.

Part 2:   **Configure the Network for SLAAC**

Step 1:   **Prepare PC-A.**

a.   Verify that the IPv6 protocol has been enabled on the Local Area Connection Properties window. If the Internet Protocol Version 6 (TCP/IPv6) check box is not checked, click to enable it.

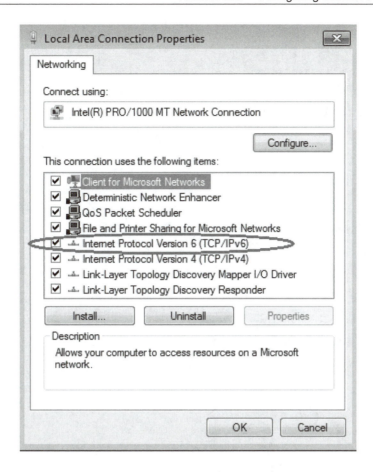

b. Start a Wireshark capture of traffic on the NIC.

c. Filter the data capture to see only RA messages. This can be done by filtering on IPv6 packets with a destination address of FF02::1, which is the all-unicast client group address. The filter entry used with Wireshark is **ipv6.dst==ff02::1**, as shown here.

Step 2: Configure R1.

a. Enable IPv6 unicast routing.

b. Assign the IPv6 unicast address to interface G0/1 according to the Addressing Table.

c. Assign FE80::1 as the IPv6 link-local address for interface G0/1.

d. Activate interface G0/1.

Step 3: Verify that R1 is part of the all-router multicast group.

Use the **show ipv6 interface g0/1** command to verify that G0/1 is part of the All-router multicast group (FF02::2). RA messages are not sent out G0/1 without that group assignment.

```
R1# show ipv6 interface g0/1
GigabitEthernet0/1 is up, line protocol is up
```

```
          IPv6 is enabled, link-local address is FE80::1
          No Virtual link-local address(es):
          Global unicast address(es):
             2001:DB8:ACAD:A::1, subnet is 2001:DB8:ACAD:A::/64
          Joined group address(es):
             FF02::1
             FF02::2
             FF02::1:FF00:1
          MTU is 1500 bytes
          ICMP error messages limited to one every 100 milliseconds
          ICMP redirects are enabled
          ICMP unreachables are sent
          ND DAD is enabled, number of DAD attempts: 1
          ND reachable time is 30000 milliseconds (using 30000)
          ND advertised reachable time is 0 (unspecified)
          ND advertised retransmit interval is 0 (unspecified)
          ND router advertisements are sent every 200 seconds
          ND router advertisements live for 1800 seconds
          ND advertised default router preference is Medium
          Hosts use stateless autoconfig for addresses.
```

Step 4: **Configure S1.**

Use the **ipv6 address autoconfig** command on VLAN 1 to obtain an IPv6 address through SLAAC.

```
     S1(config)# interface vlan 1
     S1(config-if)# ipv6 address autoconfig
     S1(config-if)# end
```

Step 5: **Verify that SLAAC provided a unicast address to S1.**

Use the **show ipv6 interface** command to verify that SLAAC provided a unicast address to VLAN1 on S1.

```
     S1# show ipv6 interface
     Vlan1 is up, line protocol is up
        IPv6 is enabled, link-local address is FE80::ED9:96FF:FEE8:8A40
        No Virtual link-local address(es):
        Stateless address autoconfig enabled
        Global unicast address(es):
           2001:DB8:ACAD:A:ED9:96FF:FEE8:8A40, subnet is 2001:DB8:ACAD:A::/64 [EUI/CAL/PRE]
              valid lifetime 2591988 preferred lifetime 604788
        Joined group address(es):
           FF02::1
           FF02::1:FFE8:8A40
        MTU is 1500 bytes
```

```
ICMP error messages limited to one every 100 milliseconds

ICMP redirects are enabled

ICMP unreachables are sent

Output features: Check hwidb

ND DAD is enabled, number of DAD attempts: 1

ND reachable time is 30000 milliseconds (using 30000)

ND NS retransmit interval is 1000 milliseconds

Default router is FE80::1 on Vlan1
```

Step 6:   **Verify that SLAAC provided IPv6 address information on PC-A.**

a.   From a command prompt on PC-A, issue the **ipconfig /all** command. Verify that PC-A is showing an IPv6 address with the 2001:db8:acad:a::/64 prefix. The Default Gateway should have the FE80::1 address.

```
Ethernet adapter Local Area Connection:

   Connection-specific DNS Suffix  . :
   Description . . . . . . . . . . . : Intel(R) PRO/1000 MT Network Connection
   Physical Address. . . . . . . . . : 00-50-56-BE-76-8C
   DHCP Enabled. . . . . . . . . . . : Yes
   Autoconfiguration Enabled . . . . : Yes
   IPv6 Address. . . . . . . . . . . : 2001:db8:acad:a:24ba:a0a0:9f0:ff88(Prefer
red)
   Temporary IPv6 Address. . . . . . : 2001:db8:acad:a:c05b:d3f7:31be:100e(Prefe
rred)
   Link-local IPv6 Address . . . . . : fe80::24ba:a0a0:9f0:ff88%11(Preferred)
   Autoconfiguration IPv4 Address. . : 169.254.255.136(Preferred)
   Subnet Mask . . . . . . . . . . . : 255.255.0.0
   Default Gateway . . . . . . . . . : fe80::1%11
   DNS Servers . . . . . . . . . . . : fec0:0:0:ffff::1%1
                                       fec0:0:0:ffff::2%1
                                       fec0:0:0:ffff::3%1
   NetBIOS over Tcpip. . . . . . . . : Enabled
```

b.   From Wireshark, look at one of the RA messages that were captured. Expand the Internet Control Message Protocol v6 layer to view the Flags and Prefix information. The first two flags control DHCPv6 usage and are not set if DHCPv6 is not configured. The prefix information is also contained within this RA message.

```
Filter: ipv6.dst==ff02::1                              ▼  Expression...  Clear  Apply
No.      Time         Source          Destination        Protocol Length Info
         3546 3615.20390 fe80::1       ff02::1            ICMPv6   118 Router Advertisement from d4:8c:b5:ce:a0:c1
3518 3972.07973 fe80::1                ff02::1            ICMPv6   118 Router Advertisement from d4:8c:b5:ce:a0:c1
3673 4130.43155 fe80::1                ff02::1            ICMPv6   118 Router Advertisement from d4:8c:b5:ce:a0:c1
3840 4284.68370 fe80::1                ff02::1            ICMPv6   118 Router Advertisement from d4:8c:b5:ce:a0:c1
3989 4435.87602 fe80::1                ff02::1            ICMPv6   118 Router Advertisement from d4:8c:b5:ce:a0:c1

⊞ Frame 3518: 118 bytes on wire (944 bits), 118 bytes captured (944 bits)
⊞ Ethernet II, Src: d4:8c:b5:ce:a0:c1 (d4:8c:b5:ce:a0:c1), Dst: IPv6mcast_00:00:00:01 (33:33:00:00:00:01)
⊞ Internet Protocol Version 6, Src: fe80::1 (fe80::1), Dst: ff02::1 (ff02::1)
⊟ Internet Control Message Protocol v6
    Type: Router Advertisement (134)
    Code: 0
    Checksum: 0x1816 [correct]
    Cur hop limit: 64
  ⊟ Flags: 0x00
      0... .... = Managed address configuration: Not set
      .0.. .... = Other configuration: Not set
      ..0. .... = Home Agent: Not set
      ...0 0... = Prf (Default Router Preference): Medium (0)
      .... .0.. = Proxy: Not set
      .... ..0. = Reserved: 0
    Router lifetime (s): 1800
    Reachable time (ms): 0
    Retrans timer (ms): 0
  ⊞ ICMPv6 Option (Source link-layer address : d4:8c:b5:ce:a0:c1)
  ⊞ ICMPv6 Option (MTU : 1500)
  ⊟ ICMPv6 Option (Prefix information : 2001:db8:acad:a::/64)
      Type: Prefix information (3)
      Length: 4 (32 bytes)
      Prefix Length: 64
    ⊞ Flag: 0xc0
      Valid Lifetime: 2592000
      Preferred Lifetime: 604800
      Reserved
      Prefix: 2001:db8:acad:a:: (2001:db8:acad:a::)
```

Part 3: Configure the Network for Stateless DHCPv6

Step 1: Configure an IPv6 DHCP server on R1.

a. Create an IPv6 DHCP pool.

```
R1(config)# ipv6 dhcp pool IPV6POOL-A
```

b. Assign a domain name to the pool.

```
R1(config-dhcpv6)# domain-name ccna-statelessDHCPv6.com
```

c. Assign a DNS server address.

```
R1(config-dhcpv6)# dns-server 2001:db8:acad:a::abcd
R1(config-dhcpv6)# exit
```

d. Assign the DHCPv6 pool to the interface.

```
R1(config)# interface g0/1
R1(config-if)# ipv6 dhcp server IPV6POOL-A
```

e. Set the DHCPv6 network discovery (ND) **other-config-flag**.

```
R1(config-if)# ipv6 nd other-config-flag
R1(config-if)# end
```

Step 2: Verify DHCPv6 settings on interface G0/1 on R1.

Use the **show ipv6 interface g0/1** command to verify that the interface is now part of the IPv6 multicast all-DHCPv6-servers group (FF02::1:2). The last line of the output from this **show** command verifies that the other-config-flag has been set.

```
R1# show ipv6 interface g0/1
GigabitEthernet0/1 is up, line protocol is up
  IPv6 is enabled, link-local address is FE80::1
  No Virtual link-local address(es):
  Global unicast address(es):
    2001:DB8:ACAD:A::1, subnet is 2001:DB8:ACAD:A::/64
  Joined group address(es):
    FF02::1
    FF02::2
    FF02::1:2
    FF02::1:FF00:1
    FF05::1:3
  MTU is 1500 bytes
  ICMP error messages limited to one every 100 milliseconds
  ICMP redirects are enabled
  ICMP unreachables are sent
  ND DAD is enabled, number of DAD attempts: 1
  ND reachable time is 30000 milliseconds (using 30000)
  ND advertised reachable time is 0 (unspecified)
  ND advertised retransmit interval is 0 (unspecified)
  ND router advertisements are sent every 200 seconds
  ND router advertisements live for 1800 seconds
  ND advertised default router preference is Medium
  Hosts use stateless autoconfig for addresses.
  Hosts use DHCP to obtain other configuration.
```

Step 3: View network changes to PC-A.

Use the **ipconfig /all** command to review the network changes. Notice that additional information, including the domain name and DNS server information, has been retrieved from the DHCPv6 server. However, the IPv6 global unicast and link-local addresses were obtained previously from SLAAC.

```
Ethernet adapter Local Area Connection:

    Connection-specific DNS Suffix  . : ccna-statelessDHCPv6.com
    Description . . . . . . . . . . . : Intel(R) PRO/1000 MT Network Connection
    Physical Address. . . . . . . . . : 00-50-56-BE-76-8C
    DHCP Enabled. . . . . . . . . . . : Yes
    Autoconfiguration Enabled . . . . : Yes
    IPv6 Address. . . . . . . . . . . : 2001:db8:acad:a:24ba:a0a0:9f0:ff88(Prefer
red)
    Temporary IPv6 Address. . . . . . : 2001:db8:acad:a:103a:4344:4b5e:ab1d(Prefe
rred)
    Link-local IPv6 Address . . . . . : fe80::24ba:a0a0:9f0:ff88%11(Preferred)
    Autoconfiguration IPv4 Address. . : 169.254.255.136(Preferred)
    Subnet Mask . . . . . . . . . . . : 255.255.0.0
    Default Gateway . . . . . . . . . : fe80::1%11
    DHCPv6 IAID . . . . . . . . . . . : 234884137
    DHCPv6 Client DUID. . . . . . . . : 00-01-00-01-17-F6-72-3D-00-0C-29-8D-54-44

    DNS Servers . . . . . . . . . . . : 2001:db8:acad:a::abcd
    NetBIOS over Tcpip. . . . . . . . : Enabled
    Connection-specific DNS Suffix Search List :
                                        ccna-statelessDHCPv6.com

Tunnel adapter isatap.{E2FC1866-B195-460A-BF40-F04F42A38FFE}:

    Media State . . . . . . . . . . . : Media disconnected
    Connection-specific DNS Suffix  . : ccna-statelessDHCPv6.com
    Description . . . . . . . . . . . : Microsoft ISATAP Adapter
    Physical Address. . . . . . . . . : 00-00-00-00-00-00-00-E0
    DHCP Enabled. . . . . . . . . . . : No
    Autoconfiguration Enabled . . . . : Yes
```

Step 4:   **View the RA messages in Wireshark.**

Scroll down to the last RA message that is displayed in Wireshark and expand it to view the ICMPv6 flag settings. Notice that the other configuration flag is set to 1.

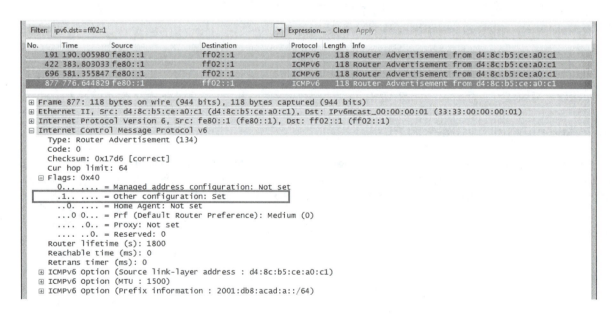

Step 5:   **Verify that PC-A did not obtain its IPv6 address from a DHCPv6 server.**

Use the **show ipv6 dhcp binding** and **show ipv6 dhcp pool** commands to verify that PC-A did not obtain an IPv6 address from the DHCPv6 pool.

```
R1# show ipv6 dhcp binding

R1# show ipv6 dhcp pool
DHCPv6 pool: IPV6POOL-A
  DNS server: 2001:DB8:ACAD:A::ABCD
  Domain name: ccna-statelessDHCPv6.com
  Active clients: 0
```

Step 6:   **Reset PC-A IPv6 network settings.**

a.   Shut down interface F0/6 on S1.

Note: Shutting down the interface F0/6 prevents PC-A from receiving a new IPv6 address before you reconfigure R1 for Stateful DHCPv6 in Part 4.

```
S1(config)# interface f0/6
S1(config-if)# shutdown
```

b.   Stop Wireshark capture of traffic on the PC-A NIC.

c.   Reset the IPv6 settings on PC-A to remove the Stateless DHCPv6 settings.

1)   Open the Local Area Connection Properties window, deselect the **Internet Protocol Version 6 (TCP/IPv6)** check box, and click **OK** to accept the change.

2)   Open the Local Area Connection Properties window again, click to enable the **Internet Protocol Version 6 (TCP/IPv6)** check box, and then click **OK** to accept the change.

Part 4:   Configure the Network for Stateful DHCPv6

Step 1:   **Prepare PC-A.**

a.   Start a Wireshark capture of traffic on the NIC.

b.   Filter the data capture to see only RA messages. This can be done by filtering on IPv6 packets with a destination address of FF02::1, which is the all-unicast client group address.

| Filter: | ipv6.dst==ff02::1 | ▼ | Expression... | Clear | Apply |

Step 2:   **Change the DHCPv6 pool on R1.**

a.   Add the network prefix to the pool.

```
R1(config)# ipv6 dhcp pool IPV6POOL-A
R1(config-dhcpv6)# address prefix 2001:db8:acad:a::/64
```

b.   Change the domain name to **ccna-statefulDHCPv6.com**.

Note: You must remove the old domain name. It is not replaced by the **domain-name** command.

```
R1(config-dhcpv6)# no domain-name ccna-statelessDHCPv6.com

R1(config-dhcpv6)# domain-name ccna-StatefulDHCPv6.com

R1(config-dhcpv6)# end
```

c. Verify DHCPv6 pool settings.

```
R1# show ipv6 dhcp pool

DHCPv6 pool: IPV6POOL-A

  Address allocation prefix: 2001:DB8:ACAD:A::/64 valid 172800 preferred 86400 (0 in
use, 0 conflicts)

  DNS server: 2001:DB8:ACAD:A::ABCD

  Domain name: ccna-StatefulDHCPv6.com

  Active clients: 0
```

d. Enter debug mode to verify the Stateful DHCPv6 address assignment.

```
R1# debug ipv6 dhcp detail

  IPv6 DHCP debugging is on (detailed)
```

Step 3:   Set the flag on G0/1 for Stateful DHCPv6.

Note: Shutting down the G0/1 interface before making changes ensures that an RA message is sent when the interface is activated.

```
R1(config)# interface g0/1

R1(config-if)# shutdown

R1(config-if)# ipv6 nd managed-config-flag

R1(config-if)# no shutdown

R1(config-if)# end
```

Step 4:   Enable interface F0/6 on S1.

Now that R1 has been configured for Stateful DHCPv6, you can reconnect PC-A to the network by activating interface F0/6 on S1.

```
S1(config)# interface f0/6

S1(config-if)# no shutdown

S1(config-if)# end
```

Step 5:   Verify Stateful DHCPv6 settings on R1.

a. Issue the **show ipv6 interface g0/1** command to verify that the interface is in Stateful DHCPv6 mode.

```
R1# show ipv6 interface g0/1

GigabitEthernet0/1 is up, line protocol is up

  IPv6 is enabled, link-local address is FE80::1

  No Virtual link-local address(es):

  Global unicast address(es):

    2001:DB8:ACAD:A::1, subnet is 2001:DB8:ACAD:A::/64
```

```
Joined group address(es):
  FF02::1
  FF02::2
  FF02::1:2
  FF02::1:FF00:1
  FF05::1:3
MTU is 1500 bytes
ICMP error messages limited to one every 100 milliseconds
ICMP redirects are enabled
ICMP unreachables are sent
ND DAD is enabled, number of DAD attempts: 1
ND reachable time is 30000 milliseconds (using 30000)
ND advertised reachable time is 0 (unspecified)
ND advertised retransmit interval is 0 (unspecified)
ND router advertisements are sent every 200 seconds
ND router advertisements live for 1800 seconds
ND advertised default router preference is Medium
Hosts use DHCP to obtain routable addresses.
Hosts use DHCP to obtain other configuration.
```

b. In a command prompt on PC-A, type **ipconfig /release6** to release the currently assigned IPv6 address. Then type **ipconfig /renew6** to request an IPv6 address from the DHCPv6 server.

c. Issue the **show ipv6 dhcp pool** command to verify the number of active clients.

```
R1# show ipv6 dhcp pool
DHCPv6 pool: IPV6POOL-A
  Address allocation prefix: 2001:DB8:ACAD:A::/64 valid 172800 preferred 86400 (1 in
use, 0 conflicts)
  DNS server: 2001:DB8:ACAD:A::ABCD
  Domain name: ccna-StatefulDHCPv6.com
  Active clients: 1
```

d. Issue the **show ipv6 dhcp binding** command to verify that PC-A received its IPv6 unicast address from the DHCP pool. Compare the client address to the link-local IPv6 address on PC-A using the **ipconfig / all** command. Compare the address provided by the **show** command to the IPv6 address listed with the **ipconfig /all** command on PC-A.

```
R1# show ipv6 dhcp binding
Client: FE80::D428:7DE2:997C:B05A
  DUID: 0001000117F6723D000C298D5444
  Username : unassigned
  IA NA: IA ID 0x0E000C29, T1 43200, T2 69120
    Address: 2001:DB8:ACAD:A:B55C:8519:8915:57CE
            preferred lifetime 86400, valid lifetime 172800
            expires at Mar 07 2013 04:09 PM (171595 seconds)
```

```
Ethernet adapter Local Area Connection:

   Connection-specific DNS Suffix  . : ccna-StatefulDHCPv6.com
   Description . . . . . . . . . . . : Intel(R) PRO/1000 MT Network Connection
   Physical Address. . . . . . . . . : 00-50-56-BE-6C-89
   DHCP Enabled. . . . . . . . . . . : Yes
   Autoconfiguration Enabled . . . . : Yes
   IPv6 Address. . . . . . . . . . . : 2001:db8:acad:a:b55c:8519:8915:57ce(Prefe
rred)
   Lease Obtained. . . . . . . . . . : Tuesday, March 05, 2013 11:53:11 AM
   Lease Expires . . . . . . . . . . : Thursday, March 07, 2013 11:53:11 AM
   IPv6 Address. . . . . . . . . . . : 2001:db8:acad:a:d428:7de2:997c:b05a(Prefe
rred)
   Temporary IPv6 Address. . . . . . : 2001:db8:acad:a:dd37:1e42:948c:225b(Prefe
rred)
   Link-local IPv6 Address . . . . . : fe80::d428:7de2:997c:b05a%11(Preferred)
   Autoconfiguration IPv4 Address. . : 169.254.176.28(Preferred)
   Subnet Mask . . . . . . . . . . . : 255.255.0.0
   Default Gateway . . . . . . . . . : fe80::1%11
   DHCPv6 IAID . . . . . . . . . . . : 234884137
   DHCPv6 Client DUID. . . . . . . . : 00-01-00-01-17-F6-72-3D-00-0C-29-8D-54-44

   DNS Servers . . . . . . . . . . . : 2001:db8:acad:a::abcd
   NetBIOS over Tcpip. . . . . . . . : Enabled
   Connection-specific DNS Suffix Search List :
                                       ccna-StatefulDHCPv6.com
```

e. Issue the **undebug all** command on R1 to stop debugging DHCPv6.

Note: Typing **u all** is the shortest form of this command and is useful to know if you are trying to stop debug messages from continually scrolling down your terminal session screen. If multiple debugs are in process, the **undebug all** command stops all of them.

```
R1# u all
```
```
All possible debugging has been turned off
```

f. Review the debug messages that appeared on your R1 terminal screen.

1) Examine the solicit message from PC-A requesting network information.

```
*Mar  5 16:42:39.775: IPv6 DHCP: Received SOLICIT from FE80::D428:7DE2:997C:B05A on
GigabitEthernet0/1

*Mar  5 16:42:39.775: IPv6 DHCP: detailed packet contents

*Mar  5 16:42:39.775:    src FE80::D428:7DE2:997C:B05A (GigabitEthernet0/1)

*Mar  5 16:42:39.775:    dst FF02::1:2

*Mar  5 16:42:39.775:    type SOLICIT(1), xid 1039238

*Mar  5 16:42:39.775:    option ELAPSED-TIME(8), len 2

*Mar  5 16:42:39.775:      elapsed-time 6300

*Mar  5 16:42:39.775:    option CLIENTID(1), len 14
```

2) Examine the reply message sent back to PC-A with the DHCP network information.

```
*Mar  5 16:42:39.779: IPv6 DHCP: Sending REPLY to FE80::D428:7DE2:997C:B05A on
GigabitEthernet0/1

*Mar  5 16:42:39.779: IPv6 DHCP: detailed packet contents

*Mar  5 16:42:39.779:    src FE80::1

*Mar  5 16:42:39.779:    dst FE80::D428:7DE2:997C:B05A (GigabitEthernet0/1)

*Mar  5 16:42:39.779:    type REPLY(7), xid 1039238

*Mar  5 16:42:39.779:    option SERVERID(2), len 10

*Mar  5 16:42:39.779:      00030001FC994775C3E0

*Mar  5 16:42:39.779:    option CLIENTID(1), len 14
```

```
*Mar  5 16:42:39.779:       00010001

R1#17F6723D000C298D5444

*Mar  5 16:42:39.779:    option IA-NA(3), len 40

*Mar  5 16:42:39.779:      IAID 0x0E000C29, T1 43200, T2 69120

*Mar  5 16:42:39.779:      option IAADDR(5), len 24

*Mar  5 16:42:39.779:        IPv6 address 2001:DB8:ACAD:A:B55C:8519:8915:57CE

*Mar  5 16:42:39.779:         preferred 86400, valid 172800

*Mar  5 16:42:39.779:    option DNS-SERVERS(23), len 16

*Mar  5 16:42:39.779:      2001:DB8:ACAD:A::ABCD

*Mar  5 16:42:39.779:    option DOMAIN-LIST(24), len 26

*Mar  5 16:42:39.779:      ccna-StatefulDHCPv6.com
```

Step 6:   Verify Stateful DHCPv6 on PC-A

a. Stop the Wireshark capture on PC-A.

b. Expand the most recent RA message listed in Wireshark. Verify that the **Managed address configuration** flag has been set.

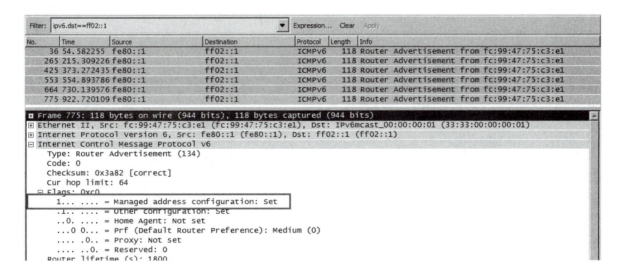

c. Change the filter in Wireshark to view **DHCPv6** packets only by typing **dhcpv6**, and then **Apply** the filter. Highlight the last DHCPv6 reply listed and expand the DHCPv6 information. Examine the DHCPv6 network information that is contained in this packet.

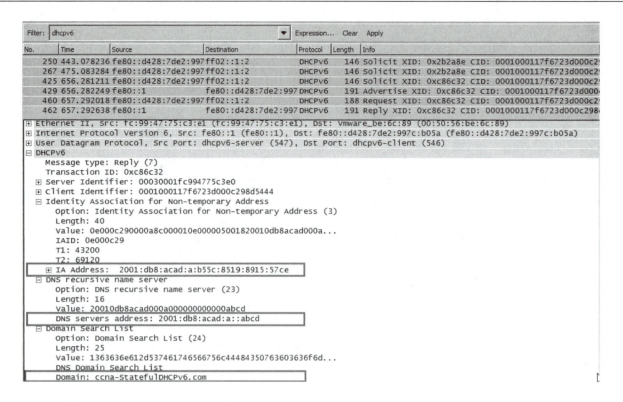

Reflection

1. What IPv6 addressing method uses more memory resources on the router configured as a DHCPv6 server, Stateless DHCPv6 or Stateful DHCPv6? Why?

_____

_____

_____

2. Which type of dynamic IPv6 address assignment is recommended by Cisco, Stateless DHCPv6 or Stateful DHCPv6?

_____

_____

Router Interface Summary Table

Router Interface Summary				
Router Model	**Ethernet Interface #1**	**Ethernet Interface #2**	**Serial Interface #1**	**Serial Interface #2**
1800	Fast Ethernet 0/0 (F0/0)	Fast Ethernet 0/1 (F0/1)	Serial 0/0/0 (S0/0/0)	Serial 0/0/1 (S0/0/1)
1900	Gigabit Ethernet 0/0 (G0/0)	Gigabit Ethernet 0/1 (G0/1)	Serial 0/0/0 (S0/0/0)	Serial 0/0/1 (S0/0/1)
2801	Fast Ethernet 0/0 (F0/0)	Fast Ethernet 0/1 (F0/1)	Serial 0/1/0 (S0/1/0)	Serial 0/1/1 (S0/1/1)
2811	Fast Ethernet 0/0 (F0/0)	Fast Ethernet 0/1 (F0/1)	Serial 0/0/0 (S0/0/0)	Serial 0/0/1 (S0/0/1)
2900	Gigabit Ethernet 0/0 (G0/0)	Gigabit Ethernet 0/1 (G0/1)	Serial 0/0/0 (S0/0/0)	Serial 0/0/1 (S0/0/1)

Note: To find out how the router is configured, look at the interfaces to identify the type of router and how many interfaces the router has. There is no way to effectively list all the combinations of configurations for each router class. This table includes identifiers for the possible combinations of Ethernet and Serial interfaces in the device. The table does not include any other type of interface, even though a specific router may contain one. An example of this might be an ISDN BRI interface. The string in parenthesis is the legal abbreviation that can be used in Cisco IOS commands to represent the interface.

10.2.4.4 Lab - Troubleshooting DHCPv6

Topology

Addressing Table

Device	Interface	IPv6 Address	Prefix Length	Default Gateway
R1	G0/1	2001:DB8:ACAD:A::1	64	N/A
S1	VLAN 1	Assigned by SLAAC	64	Assigned by SLAAC
PC-A	NIC	Assigned by SLAAC and DHCPv6	64	Assigned by SLAAC

Objectives

Part 1: Build the Network and Configure Basic Device Settings

Part 2: Troubleshoot IPv6 Connectivity

Part 3: Troubleshoot Stateless DHCPv6

Background / Scenario

The ability to troubleshoot network issues is a very useful skill for network administrators. It is important to understand IPv6 address groups and how they are used when troubleshooting a network. Knowing what commands to use to extract IPv6 network information is necessary to effectively troubleshoot.

In this lab, you will load configurations on R1 and S1. These configurations will contain issues that prevent Stateless DHCPv6 from functioning on the network. You will troubleshoot R1 and S1 to resolve these issues.

Note: The routers used with CCNA hands-on labs are Cisco 1941 Integrated Services Routers (ISRs) with Cisco IOS Release 15.2(4)M3 (universalk9 image). The switches used are Cisco Catalyst 2960s with Cisco IOS Release 15.0(2) (lanbasek9 image). Other routers, switches and Cisco IOS versions can be used. Depending on the model and Cisco IOS version, the commands available and output produced might vary from what is shown in the labs. Refer to the Router Interface Summary Table at the end of this lab for the correct interface identifiers.

Note: Make sure that the router and switch have been erased and have no startup configurations. If you are unsure, contact your instructor.

Note: The default bias template used by the Switch Database Manager (SDM) does not provide IPv6 address capabilities. Verify that SDM is using either the **dual-ipv4-and-ipv6** template or the **lanbase-routing** template. The new template will be used after reboot even if the configuration is not saved.

```
S1# show sdm prefer
```

Follow this configuration to assign the **dual-ipv4-and-ipv6** template as the default SDM template:

```
S1# config t
S1(config)# sdm prefer dual-ipv4-and-ipv6 default
S1(config)# end
S1# reload
```

Required Resources

- 1 Router (Cisco 1941 with Cisco IOS Release 15.2(4)M3 universal image or comparable)
- 1 Switch (Cisco 2960 with Cisco IOS Release 15.0(2) lanbasek9 image or comparable)
- 1 PC (Windows 7, Vista, or XP with terminal emulation program, such as Tera Term)
- Console cables to configure the Cisco IOS devices via the console ports
- Ethernet cables as shown in the topology

Part 1: Build the Network and Configure Basic Device Settings

In Part 1, you will set up the network topology and clear any configurations if necessary. You will configure basic settings on the router and switch. Then you will load the provided IPv6 configurations before you start troubleshooting.

Step 1: Cable the network as shown in the topology.

Step 2: Initialize and reload the router and the switch.

Step 3: Configure basic settings on the router and switch.

a. Disable DNS lookup.

b. Configure device names as shown in the topology.

c. Encrypt plain text passwords.

d. Create a MOTD banner warning users that unauthorized access is prohibited.

e. Assign **class** as the encrypted privileged EXEC mode password.

f. Assign **cisco** as the console and vty passwords and enable login.

g. Configure **logging synchronous** to prevent console messages from interrupting command entry.

Step 4: Load the IPv6 configuration to R1.

```
ip domain name ccna-lab.com


ipv6 dhcp pool IPV6POOL-A
 dns-server 2001:DB8:ACAD:CAFE::A
 domain-name ccna-lab.com
interface g0/0
```

```
  no ip address
  shutdown
  duplex auto
  speed auto
interface g0/1
  no ip address
  duplex auto
  speed auto
  ipv6 address FE80::1 link-local
  ipv6 address 2001:DB8:ACAD:A::11/64




end
```

Step 5: Load the IPv6 configuration to S1.

```
interface range f0/1-24
  shutdown



interface range g0/1-2
  shutdown
interface Vlan1
  shutdown



end
```

Step 6: Save the running configurations on R1 and S1.

Step 7: Verify that IPv6 is enabled on PC-A.

Verify that IPv6 has been enabled in the Local Area Connection Properties window on PC-A.

Part 2: Troubleshoot IPv6 Connectivity

In Part 2, you will test and verify Layer 3 IPv6 connectivity on the network. Continue troubleshooting the network until Layer 3 connectivity has been established on all devices. Do not continue to Part 3 until you have successfully completed Part 2.

Step 1: **Troubleshoot IPv6 interfaces on R1.**

a. According to the topology, which interface must be active on R1 for network connectivity to be established? Record any commands used to identify which interfaces are active.

_____

b. If necessary, take the steps required to bring up the interface. Record the commands used to correct the configuration errors and verify that the interface is active.

_____

_____

c. Identify the IPv6 addresses configured on R1. Record the addresses found and the commands used to view the IPv6 addresses.

_____

_____

d. Determine if a configuration error has been made. If any errors are identified, record all the commands used to correct the configuration.

_____

_____

_____

e. On R1, what multicast group is needed for SLAAC to function?

_____

f. What command is used to verify that R1 is a member of that group?

_____

g. If R1 is not a member of the multicast group that is needed for SLAAC to function correctly, make the necessary changes to the configuration so that it joins the group. Record any commands necessary to correct the configurations errors.

_____

h. Re-issue the command to verify that interface G0/1 has joined the all-routers multicast group (FF02::2).

Note: If you are unable to join the all-routers multicast group, you may need to save your current configuration and reload the router.

Step 2: **Troubleshoot S1.**

a. Are the interfaces needed for network connectivity active on S1? _____

Record any commands that are used to activate necessary interfaces on S1.

_____

_____

_____

_____

b. What command could you use to determine if an IPv6 unicast address has been assigned to S1?

_____

c. Does S1 have an IPv6 unicast address configured? If so, what is it?

_____

d. If S1 is not receiving a SLAAC address, make the necessary configuration changes to allow it to receive one. Record the commands used.

_____

_____

e. Re-issue the command that verifies that the interface now receives a SLAAC address.

f. Can S1 ping the IPv6 unicast address assigned to the G0/1 interface assigned to R1?

_____

Step 3:   **Troubleshoot PC-A.**

a. Issue the command used on PC-A to verify the IPv6 address assigned. Record the command.

_____

b. What is the IPv6 unicast address SLAAC is providing to PC-A?

_____

c. Can PC-A ping the default gateway address that was assigned by SLAAC?

_____

d. Can PC-A ping the management interface on S1?

_____

_____

_____

Note: Continue troubleshooting until you can ping R1 and S1 from PC-A.

Part 3:   **Troubleshoot Stateless DHCPv6**

In Part 3, you will test and verify that Stateless DHCPv6 is working correctly on the network. You will need to use the correct IPv6 CLI commands on the router to determine if Stateless DHCPv6 is working. You may want to use debug to help determine if the DHCP server is being solicited.

Step 1:   **Determine if Stateless DHCPv6 is functioning correctly.**

a. What is the name of the IPv6 DHCP pool? How did you determine this?

_____

_____

b. What network information is listed in the DHCPv6 pool?

_____

_____

c. Was the DHCPv6 information assigned to PC-A? How did you determine this?

_____

_____

Step 2: **Troubleshoot R1.**

a.   What commands can be used to determine if R1 is configured for Stateless DHCPv6?

_____

_____

b.   Is the G0/1 interface on R1 in Stateless DHCPv6 mode?

_____

_____

c.   What command can be used to have R1 join the all-DHCPv6 server group?

_____

_____

d.   Verify that the all-DHCPv6 server group is configured for interface G0/1.

e.   Will PC-A receive the DHCP information now? Explain?

_____

_____

f.   What is missing from the configuration of G0/1 that causes hosts to use the DCHP server to retrieve other network information?

_____

g.   Reset the IPv6 settings on PC-A.

1) Open the Local Area Connection Properties window, deselect the Internet Protocol Version 6 (TCP/IPv6) check box, and then click **OK** to accept the change.

2) Open the Local Area Connection Properties window again, click the Internet Protocol Version 6 (TCP/IPv6) check box, and then click **OK** to accept the change.

h. Issue the command to verify changes have been made on PC-A.

Note: Continue troubleshooting until PC-A receives the additional DHCP information from R1.

Reflection

1. What command is needed in the DHCPv6 pool for Stateful DHCPv6 that is not needed for Stateless DHCPv6? Why?

_____

_____

_____

2. What command is needed on the interface to change the network to use Stateful DHCPv6 instead of Stateless DHCPv6?

_____

_____

Router Interface Summary Table

Router Interface Summary				
Router Model	**Ethernet Interface #1**	**Ethernet Interface #2**	**Serial Interface #1**	**Serial Interface #2**
1800	Fast Ethernet 0/0 (F0/0)	Fast Ethernet 0/1 (F0/1)	Serial 0/0/0 (S0/0/0)	Serial 0/0/1 (S0/0/1)
1900	Gigabit Ethernet 0/0 (G0/0)	Gigabit Ethernet 0/1 (G0/1)	Serial 0/0/0 (S0/0/0)	Serial 0/0/1 (S0/0/1)
2801	Fast Ethernet 0/0 (F0/0)	Fast Ethernet 0/1 (F0/1)	Serial 0/1/0 (S0/1/0)	Serial 0/1/1 (S0/1/1)
2811	Fast Ethernet 0/0 (F0/0)	Fast Ethernet 0/1 (F0/1)	Serial 0/0/0 (S0/0/0)	Serial 0/0/1 (S0/0/1)
2900	Gigabit Ethernet 0/0 (G0/0)	Gigabit Ethernet 0/1 (G0/1)	Serial 0/0/0 (S0/0/0)	Serial 0/0/1 (S0/0/1)
Note: To find out how the router is configured, look at the interfaces to identify the type of router and how many interfaces the router has. There is no way to effectively list all the combinations of configurations for each router class. This table includes identifiers for the possible combinations of Ethernet and Serial interfaces in the device. The table does not include any other type of interface, even though a specific router may contain one. An example of this might be an ISDN BRI interface. The string in parenthesis is the legal abbreviation that can be used in Cisco IOS commands to represent the interface.				

10.3.1.1 Class Activity – IoE and DHCP

Objective

Configure DHCP for IPv4 or IPv6 on a Cisco 1941 router.

Scenario

This chapter presents the concept of using the DHCP process in a small- to medium-sized business network; however, DHCP also has other uses!

With the advent of the Internet of Everything (IoE), any device in your home capable of wired or wireless connectivity to a network will be able to be accessed from just about anywhere.

Using Packet Tracer for this modeling activity, perform the following tasks:

- Configure a Cisco 1941 router (or DHCP-server-capable ISR device) for IPv4 or IPv6 DHCP addressing.
- Think of five devices in your home you would like to receive IP addresses from the router's DHCP service. Set the end devices to claim DHCP addresses from the DHCP server.
- Show output validating that each end device secures an IP address from the server. Save your output information via a screen capture program or use the **PrtScrn** key command.
- Present your findings to a fellow classmate or to the class.

Required Resources

Packet Tracer software

Reflection

1. Why would a user want to use a Cisco 1941 router to configure DHCP on his home network? Wouldn't a smaller ISR be good enough to use as a DHCP server?

_____

2. How do you think small- medium-sized businesses are able to use DHCP IP address allocation in the IoE and IPv6 network world? Brainstorm and record five possible answers.

_____

Chapter 11 — Network Address Translation for IPv4

11.0.1.2 Class Activity – Conceptual NAT

Objective

Describe NAT characteristics.

Scenario

You work for a large university or school system. Because you are the network administrator, many professors, administrative workers, and other network administrators need your assistance with their networks on a daily basis. They call you at all working hours of the day and, because of the number of telephone calls, you cannot complete your regular network administration tasks.

You need to find a way to limit when you take calls and from whom. You also need to mask your telephone number so that when you call someone, another number is displayed to the recipient.

This scenario describes a very common problem for most small- to medium-sized businesses. Visit, "How Network Address Translation Works" located at http://computer.howstuffworks.com/nat.htm/printable to view more information about how the digital world handles these types of workday interruptions.

Use the PDF provided accompanying this activity to reflect further on how a process, known as NAT, could be the answer to this scenario's challenge.

Resources

Internet connection

Directions

Step 1:   **Read Information on the Internet Site.**

    a.   Go to "How Network Address Translation Works" located at http://computer.howstuffworks.com/nat.htm/printable

    b.   Read the information provided to introduce the basic concepts of NAT.

    c.   Record five facts you find to be interesting about the NAT process.

Step 2:   **View the NAT graphics.**

    a.   On the same Internet page, look at the types of NAT that are available for configuration on most networks.

    b.   Define the four NAT types:

        1)   Static NAT

        2)   Dynamic NAT

        3)   NAT Overload

        4)   NAT Overlap

Step 3:   **Meet together in a full-class setting.**

    a.   Report your five NAT facts to the class.

    b.   As other students state their interesting facts to the class, check off the stated fact if you already recorded it.

    c.   If a student reports a fact to the class that you did not record, add it to your list.

11.2.2.6 Lab – Configuring Dynamic and Static NAT

Topology

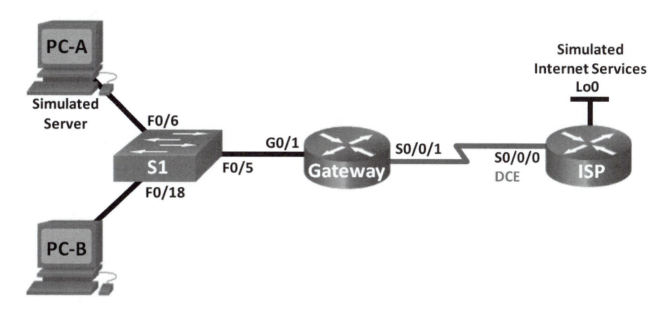

Addressing Table

Device	Interface	IP Address	Subnet Mask	Default Gateway
Gateway	G0/1	192.168.1.1	255.255.255.0	N/A
	S0/0/1	209.165.201.18	255.255.255.252	N/A
ISP	S0/0/0 (DCE)	209.165.201.17	255.255.255.252	N/A
	Lo0	192.31.7.1	255.255.255.255	N/A
PC-A (Simulated Server)	NIC	192.168.1.20	255.255.255.0	192.168.1.1
PC-B	NIC	192.168.1.21	255.255.255.0	192.168.1.1

Objectives

Part 1: Build the Network and Verify Connectivity

Part 2: Configure and Verify Static NAT

Part 3: Configure and Verify Dynamic NAT

Background / Scenario

Network Address Translation (NAT) is the process where a network device, such as a Cisco router, assigns a public address to host devices inside a private network. The main reason to use NAT is to reduce the number of public IP addresses that an organization uses because the number of available IPv4 public addresses is limited.

In this lab, an ISP has allocated the public IP address space of 209.165.200.224/27 to a company. This provides the company with 30 public IP addresses. The addresses, 209.165.200.225 to 209.165.200.241, are for static allocation and 209.165.200.242 to 209.165.200.254 are for dynamic allocation. A static route is used from the ISP to the gateway router, and a default route is used from the gateway to the ISP router. The ISP connection to the Internet is simulated by a loopback address on the ISP router.

Note: The routers used with CCNA hands-on labs are Cisco 1941 Integrated Services Routers (ISRs) with Cisco IOS Release 15.2(4)M3 (universalk9 image). The switches used are Cisco Catalyst 2960s with Cisco IOS Release 15.0(2) (lanbasek9 image). Other routers, switches and Cisco IOS versions can be used. Depending on the model and Cisco IOS version, the commands available and output produced might vary from what is shown in the labs. Refer to the Router Interface Summary Table at the end of this lab for the correct interface identifiers.

Note: Make sure that the routers and switch have been erased and have no startup configurations. If you are unsure, contact your instructor.

Required Resources

- 2 Routers (Cisco 1941 with Cisco IOS Release 15.2(4)M3 universal image or comparable)
- 1 Switch (Cisco 2960 with Cisco IOS Release 15.0(2) lanbasek9 image or comparable)
- 2 PCs (Windows 7, Vista, or XP with terminal emulation program, such as Tera Term)
- Console cables to configure the Cisco IOS devices via the console ports
- Ethernet and serial cables as shown in the topology

Part 1: Build the Network and Verify Connectivity

In Part 1, you will set up the network topology and configure basic settings, such as the interface IP addresses, static routing, device access, and passwords.

Step 1: Cable the network as shown in the topology.

Attach the devices as shown in the topology diagram, and cable as necessary.

Step 2: Configure PC hosts.

Step 3: Initialize and reload the routers and switches as necessary.

Step 4: Configure basic settings for each router.

a. Disable DNS lookup.

b. Configure IP addresses for the routers as listed in the Addressing Table.

c. Set the clock rate to **1280000** for the DCE serial interfaces.

d. Configure device name as shown in the topology.

e. Assign **cisco** as the console and vty passwords.

f. Assign **class** as the encrypted privileged EXEC mode password.

g. Configure **logging synchronous** to prevent console messages from interrupting the command entry.

Step 5:   **Create a simulated web server on ISP.**

a. Create a local user named **webuser** with an encrypted password of **webpass**.

    ISP(config)# **username webuser privilege 15 secret webpass**

b. Enable the HTTP server service on ISP.

    ISP(config)# **ip http server**

c. Configure the HTTP service to use the local user database.

    ISP(config)# **ip http authentication local**

Step 6:   **Configure static routing.**

a. Create a static route from the ISP router to the Gateway router using the assigned public network address range 209.165.200.224/27.

    ISP(config)# **ip route 209.165.200.224 255.255.255.224 209.165.201.18**

b. Create a default route from the Gateway router to the ISP router.

    Gateway(config)# **ip route 0.0.0.0 0.0.0.0 209.165.201.17**

Step 7:   **Save the running configuration to the startup configuration.**

Step 8:   **Verify network connectivity.**

a. From the PC hosts, ping the G0/1 interface on the Gateway router. Troubleshoot if the pings are unsuccessful.

b. Display the routing tables on both routers to verify that the static routes are in the routing table and configured correctly on both routers.

Part 2:   **Configure and Verify Static NAT**

Static NAT uses a one-to-one mapping of local and global addresses, and these mappings remain constant. Static NAT is particularly useful for web servers or devices that must have static addresses that are accessible from the Internet.

Step 1:   **Configure a static mapping.**

A static map is configured to tell the router to translate between the private inside server address 192.168.1.20 and the public address 209.165.200.225. This allows a user from the Internet to access PC-A. PC-A is simulating a server or device with a constant address that can be accessed from the Internet.

    Gateway(config)# **ip nat inside source static 192.168.1.20 209.165.200.225**

Step 2:   **Specify the interfaces.**

Issue the **ip nat inside** and **ip nat outside** commands to the interfaces.

    Gateway(config)# **interface g0/1**

    Gateway(config-if)# **ip nat inside**

    Gateway(config-if)# **interface s0/0/1**

    Gateway(config-if)# **ip nat outside**

Step 3:   **Test the configuration.**

a.   Display the static NAT table by issuing the **show ip nat translations** command.

```
Gateway# show ip nat translations
Pro Inside global      Inside local      Outside local      Outside global
--- 209.165.200.225    192.168.1.20      ---                ---
```

What is the translation of the Inside local host address?

192.168.1.20 = _____

The Inside global address is assigned by?

_____

The Inside local address is assigned by?

_____

b.   From PC-A, ping the Lo0 interface (192.31.7.1) on ISP. If the ping was unsuccessful, troubleshoot and correct the issues. On the Gateway router, display the NAT table.

```
Gateway# show ip nat translations
Pro Inside global      Inside local      Outside local      Outside global
icmp 209.165.200.225:1 192.168.1.20:1   192.31.7.1:1       192.31.7.1:1
--- 209.165.200.225    192.168.1.20      ---                ---
```

A NAT entry was added to the table with ICMP listed as the protocol when PC-A sent an ICMP request (ping) to 192.31.7.1 on ISP.

What port number was used in this ICMP exchange? _____

Note: It may be necessary to disable the PC-A firewall for the ping to be successful.

c.   From PC-A, telnet to the ISP Lo0 interface and display the NAT table.

```
Pro Inside global         Inside local         Outside local      Outside global
icmp 209.165.200.225:1    192.168.1.20:1       192.31.7.1:1       192.31.7.1:1
tcp 209.165.200.225:1034  192.168.1.20:1034    192.31.7.1:23      192.31.7.1:23
--- 209.165.200.225       192.168.1.20         ---                ---
```

Note: The NAT for the ICMP request may have timed out and been removed from the NAT table.

What was the protocol used in this translation? _____

What are the port numbers used?

Inside global / local: _____

Outside global / local: _____

d.   Because static NAT was configured for PC-A, verify that pinging from ISP to PC-A at the static NAT public address (209.165.200.225) is successful.

e. On the Gateway router, display the NAT table to verify the translation.

```
Gateway# show ip nat translations
Pro Inside global      Inside local       Outside local      Outside global
icmp 209.165.200.225:12 192.168.1.20:12   209.165.201.17:12  209.165.201.17:12
--- 209.165.200.225    192.168.1.20       ---                ---
```

Notice that the Outside local and Outside global addresses are the same. This address is the ISP remote network source address. For the ping from the ISP to succeed, the Inside global static NAT address 209.165.200.225 was translated to the Inside local address of PC-A (192.168.1.20).

f. Verify NAT statistics by using the **show ip nat statistics** command on the Gateway router.

```
Gateway# show ip nat statics
Total active translations: 2 (1 static, 1 dynamic; 1 extended)
Peak translations: 2, occurred 00:02:12 ago
Outside interfaces:
  Serial0/0/1
Inside interfaces:
  GigabitEthernet0/1
Hits: 39  Misses: 0
CEF Translated packets: 39, CEF Punted packets: 0
Expired translations: 3
Dynamic mappings:

Total doors: 0
Appl doors: 0
Normal doors: 0
Queued Packets: 0
```

Note: This is only a sample output. Your output may not match exactly.

Part 3: Configure and Verify Dynamic NAT

Dynamic NAT uses a pool of public addresses and assigns them on a first-come, first-served basis. When an inside device requests access to an outside network, dynamic NAT assigns an available public IPv4 address from the pool. Dynamic NAT results in a many-to-many address mapping between local and global addresses.

Step 1: Clear NATs.

Before proceeding to add dynamic NATs, clear the NATs and statistics from Part 2.

```
Gateway# clear ip nat translation *
Gateway# clear ip nat statistics
```

Step 2: Define an access control list (ACL) that matches the LAN private IP address range.

ACL 1 is used to allow 192.168.1.0/24 network to be translated.

```
Gateway(config)# access-list 1 permit 192.168.1.0 0.0.0.255
```

Step 3: **Verify that the NAT interface configurations are still valid.**

Issue the **show ip nat statistics** command on the Gateway router to verify the NAT configurations.

Step 4: **Define the pool of usable public IP addresses.**

```
Gateway(config)# ip nat pool public_access 209.165.200.242 209.165.200.254
netmask 255.255.255.224
```

Step 5: **Define the NAT from the inside source list to the outside pool.**

Note: Remember that NAT pool names are case-sensitive and the pool name entered here must match that used in the previous step.

```
Gateway(config)# ip nat inside source list 1 pool public_access
```

Step 6: **Test the configuration.**

a. From PC-B, ping the Lo0 interface (192.31.7.1) on ISP. If the ping was unsuccessful, troubleshoot and correct the issues. On the Gateway router, display the NAT table.

```
Gateway# show ip nat translations
Pro Inside global      Inside local      Outside local      Outside global
--- 209.165.200.225    192.168.1.20      ---                ---
icmp 209.165.200.242:1 192.168.1.21:1    192.31.7.1:1       192.31.7.1:1
--- 209.165.200.242    192.168.1.21      ---                ---
```

What is the translation of the Inside local host address for PC-B?

192.168.1.21 = _____

A dynamic NAT entry was added to the table with ICMP as the protocol when PC-B sent an ICMP message to 192.31.7.1 on ISP.

What port number was used in this ICMP exchange? _____

b. From PC-B, open a browser and enter the IP address of the ISP-simulated web server (Lo0 interface). When prompted, log in as **webuser** with a password of **webpass**.

c. Display the NAT table.

```
Pro Inside global       Inside local       Outside local       Outside global
--- 209.165.200.225     192.168.1.20       ---                 ---
tcp 209.165.200.242:1038 192.168.1.21:1038 192.31.7.1:80       192.31.7.1:80
tcp 209.165.200.242:1039 192.168.1.21:1039 192.31.7.1:80       192.31.7.1:80
tcp 209.165.200.242:1040 192.168.1.21:1040 192.31.7.1:80       192.31.7.1:80
tcp 209.165.200.242:1041 192.168.1.21:1041 192.31.7.1:80       192.31.7.1:80
tcp 209.165.200.242:1042 192.168.1.21:1042 192.31.7.1:80       192.31.7.1:80
tcp 209.165.200.242:1043 192.168.1.21:1043 192.31.7.1:80       192.31.7.1:80
tcp 209.165.200.242:1044 192.168.1.21:1044 192.31.7.1:80       192.31.7.1:80
tcp 209.165.200.242:1045 192.168.1.21:1045 192.31.7.1:80       192.31.7.1:80
tcp 209.165.200.242:1046 192.168.1.21:1046 192.31.7.1:80       192.31.7.1:80
tcp 209.165.200.242:1047 192.168.1.21:1047 192.31.7.1:80       192.31.7.1:80
tcp 209.165.200.242:1048 192.168.1.21:1048 192.31.7.1:80       192.31.7.1:80
tcp 209.165.200.242:1049 192.168.1.21:1049 192.31.7.1:80       192.31.7.1:80
tcp 209.165.200.242:1050 192.168.1.21:1050 192.31.7.1:80       192.31.7.1:80
tcp 209.165.200.242:1051 192.168.1.21:1051 192.31.7.1:80       192.31.7.1:80
tcp 209.165.200.242:1052 192.168.1.21:1052 192.31.7.1:80       192.31.7.1:80
--- 209.165.200.242     192.168.1.22       ---                 ---
```

What protocol was used in this translation? _____

What port numbers were used?

Inside: _____

outside: _____

What well-known port number and service was used? _____

d. Verify NAT statistics by using the **show ip nat statistics** command on the Gateway router.

```
Gateway# show ip nat statistics
Total active translations: 3 (1 static, 2 dynamic; 1 extended)
Peak translations: 17, occurred 00:06:40 ago
Outside interfaces:
  Serial0/0/1
Inside interfaces:
  GigabitEthernet0/1
Hits: 345  Misses: 0
CEF Translated packets: 345, CEF Punted packets: 0
Expired translations: 20
Dynamic mappings:
-- Inside Source
```

```
[Id: 1] access-list 1 pool public_access refcount 2
 pool public_access: netmask 255.255.255.224
        start 209.165.200.242 end 209.165.200.254
        type generic, total addresses 13, allocated 1 (7%), misses 0
```

```
Total doors: 0
Appl doors: 0
Normal doors: 0
Queued Packets: 0
```

Note: This is only a sample output. Your output may not match exactly.

Step 7:   **Remove the static NAT entry.**

In Step 7, the static NAT entry is removed and you can observe the NAT entry.

a.   Remove the static NAT from Part 2. Enter **yes** when prompted to delete child entries.

Gateway(config)# **no ip nat inside source static 192.168.1.20 209.165.200.225**

```
Static entry in use, do you want to delete child entries? [no]: yes
```

b.   Clear the NATs and statistics.

c.   Ping the ISP (192.31.7.1) from both hosts.

d.   Display the NAT table and statistics.

```
Gateway# show ip nat statistics
Total active translations: 4 (0 static, 4 dynamic; 2 extended)
Peak translations: 15, occurred 00:00:43 ago
Outside interfaces:
  Serial0/0/1
Inside interfaces:
  GigabitEthernet0/1
Hits: 16  Misses: 0
CEF Translated packets: 285, CEF Punted packets: 0
Expired translations: 11
Dynamic mappings:
-- Inside Source
[Id: 1] access-list 1 pool public_access refcount 4
 pool public_access: netmask 255.255.255.224
        start 209.165.200.242 end 209.165.200.254
        type generic, total addresses 13, allocated 2 (15%), misses 0
```

```
Total doors: 0
Appl doors: 0
```

```
        Normal doors: 0

        Queued Packets: 0


        Gateway# show ip nat translation
        Pro Inside global        Inside local       Outside local       Outside global
        icmp 209.165.200.243:512 192.168.1.20:512 192.31.7.1:512       192.31.7.1:512
        --- 209.165.200.243      192.168.1.20       ---                 ---
        icmp 209.165.200.242:512 192.168.1.21:512 192.31.7.1:512       192.31.7.1:512
        --- 209.165.200.242      192.168.1.21       ---                 ---
```

Note: This is only a sample output. Your output may not match exactly.

Reflection

1. Why would NAT be used in a network?

   _____

   _____

   _____

2. What are the limitations of NAT?

   _____

   _____

   _____

Router Interface Summary Table

Router Interface Summary				
Router Model	Ethernet Interface #1	Ethernet Interface #2	Serial Interface #1	Serial Interface #2
1800	Fast Ethernet 0/0 (F0/0)	Fast Ethernet 0/1 (F0/1)	Serial 0/0/0 (S0/0/0)	Serial 0/0/1 (S0/0/1)
1900	Gigabit Ethernet 0/0 (G0/0)	Gigabit Ethernet 0/1 (G0/1)	Serial 0/0/0 (S0/0/0)	Serial 0/0/1 (S0/0/1)
2801	Fast Ethernet 0/0 (F0/0)	Fast Ethernet 0/1 (F0/1)	Serial 0/1/0 (S0/1/0)	Serial 0/1/1 (S0/1/1)
2811	Fast Ethernet 0/0 (F0/0)	Fast Ethernet 0/1 (F0/1)	Serial 0/0/0 (S0/0/0)	Serial 0/0/1 (S0/0/1)
2900	Gigabit Ethernet 0/0 (G0/0)	Gigabit Ethernet 0/1 (G0/1)	Serial 0/0/0 (S0/0/0)	Serial 0/0/1 (S0/0/1)

Note: To find out how the router is configured, look at the interfaces to identify the type of router and how many interfaces the router has. There is no way to effectively list all the combinations of configurations for each router class. This table includes identifiers for the possible combinations of Ethernet and Serial interfaces in the device. The table does not include any other type of interface, even though a specific router may contain one. An example of this might be an ISDN BRI interface. The string in parenthesis is the legal abbreviation that can be used in Cisco IOS commands to represent the interface.

11.2.3.7 Lab – Configuring NAT Pool Overload and PAT

Topology

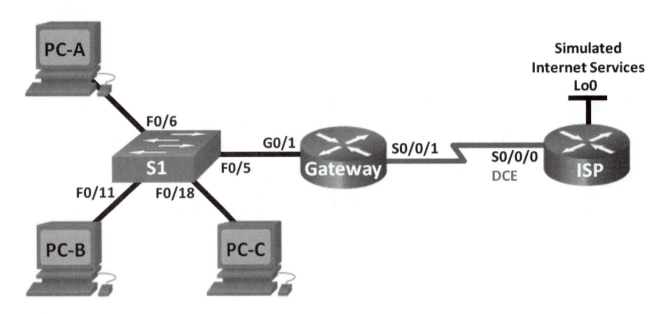

Addressing Table

Device	Interface	IP Address	Subnet Mask	Default Gateway
Gateway	G0/1	192.168.1.1	255.255.255.0	N/A
	S0/0/1	209.165.201.18	255.255.255.252	N/A
ISP	S0/0/0 (DCE)	209.165.201.17	255.255.255.252	N/A
	Lo0	192.31.7.1	255.255.255.255	N/A
PC-A	NIC	192.168.1.20	255.255.255.0	192.168.1.1
PC-B	NIC	192.168.1.21	255.255.255.0	192.168.1.1
PC-C	NIC	192.168.1.22	255.255.255.0	192.168.1.1

Objectives

Part 1: Build the Network and Verify Connectivity

Part 2: Configure and Verify NAT Pool Overload

Part 3: Configure and Verify PAT

Background / Scenario

In the first part of the lab, your company is allocated the public IP address range of 209.165.200.224/29 by the ISP. This provides the company with six public IP addresses. Dynamic NAT pool overload uses a pool of IP addresses in a many-to-many relationship. The router uses the first IP address in the pool and assigns connections using the IP address plus a unique port number. After the maximum number of translations for a single IP address have been reached on the router (platform and hardware specific), it uses the next IP address in the pool.

In Part 2, the ISP has allocated a single IP address, 209.165.201.18, to your company for use on the Internet connection from the company Gateway router to the ISP. You will use the Port Address Translation (PAT) to convert multiple internal addresses into the one usable public address. You will test, view, and verify that the translations are taking place, and you will interpret the NAT/PAT statistics to monitor the process.

Note: The routers used with CCNA hands-on labs are Cisco 1941 Integrated Services Routers (ISRs) with Cisco IOS Release 15.2(4)M3 (universalk9 image). The switches used are Cisco Catalyst 2960s with Cisco IOS Release 15.0(2) (lanbasek9 image). Other routers, switches and Cisco IOS versions can be used. Depending on the model and Cisco IOS version, the commands available and output produced might vary from what is shown in the labs. Refer to the Router Interface Summary Table at the end of this lab for the correct interface identifiers.

Note: Make sure that the routers and switch have been erased and have no startup configurations. If you are unsure, contact your instructor.

Required Resources

- 2 Routers (Cisco 1941 with Cisco IOS Release 15.2(4)M3 universal image or comparable)
- 1 Switch (Cisco 2960 with Cisco IOS Release 15.0(2) lanbasek9 image or comparable)
- 3 PCs (Windows 7, Vista, or XP with terminal emulation program, such as Tera Term)
- Console cables to configure the Cisco IOS devices via the console ports
- Ethernet and serial cables as shown in the topology

Part 1: Build the Network and Verify Connectivity

In Part 1, you will set up the network topology and configure basic settings, such as the interface IP addresses, static routing, device access, and passwords.

Step 1: Cable the network as shown in the topology.

Step 2: Configure PC hosts.

Step 3: Initialize and reload the routers and switches.

Step 4: Configure basic settings for each router.

a. Disable DNS lookup.

b. Configure IP addresses for the routers as listed in the Addressing Table.

c. Set the clock rate to **128000** for DCE serial interface.

d. Configure device name as shown in the topology.

e. Assign **cisco** as the console and vty passwords.

f. Assign **class** as the encrypted privileged EXEC mode password.

g. Configure **logging synchronous** to prevent console messages from interrupting the command entry.

Step 5:   **Configure static routing.**

a.   Create a static route from the ISP router to the Gateway router.

```
ISP(config)# ip route 209.165.200.224 255.255.255.248 209.165.201.18
```

b.   Create a default route from the Gateway router to the ISP router.

```
Gateway(config)# ip route 0.0.0.0 0.0.0.0 209.165.201.17
```

Step 6:   **Verify network connectivity.**

a.   From the PC hosts, ping the G0/1 interface on the Gateway router. Troubleshoot if the pings are unsuccessful.

b.   Verify that the static routes are configured correctly on both routers.

Part 2:   Configure and Verify NAT Pool Overload

In Part 2, you will configure the Gateway router to translate the IP addresses from the 192.168.1.0/24 network to one of the six usable addresses in the 209.165.200.224/29 range.

Step 1:   **Define an access control list that matches the LAN private IP addresses.**

ACL 1 is used to allow the 192.168.1.0/24 network to be translated.

```
Gateway(config)# access-list 1 permit 192.168.1.0 0.0.0.255
```

Step 2:   **Define the pool of usable public IP addresses.**

```
Gateway(config)# ip nat pool public_access 209.165.200.225  209.165.200.230
netmask 255.255.255.248
```

Step 3:   **Define the NAT from the inside source list to the outside pool.**

```
Gateway(config)# ip nat inside source list 1 pool public_access overload
```

Step 4:   **Specify the interfaces.**

Issue the **ip nat inside** and **ip nat outside** commands to the interfaces.

```
Gateway(config)# interface g0/1
Gateway(config-if)# ip nat inside
Gateway(config-if)# interface s0/0/1
Gateway(config-if)# ip nat outside
```

Step 5:   **Verify the NAT pool overload configuration.**

a.   From each PC host, ping the 192.31.7.1 address on the ISP router.

b.   Display NAT statistics on the Gateway router.

```
Gateway# show ip nat statistics
Total active translations: 3 (0 static, 3 dynamic; 3 extended)
Peak translations: 3, occurred 00:00:25 ago
```

```
Outside interfaces:
  Serial0/0/1
Inside interfaces:
  GigabitEthernet0/1
Hits: 24  Misses: 0
CEF Translated packets: 24, CEF Punted packets: 0
Expired translations: 0
Dynamic mappings:
-- Inside Source
[Id: 1] access-list 1 pool public_access refcount 3
 pool public_access: netmask 255.255.255.248
        start 209.165.200.225 end 209.165.200.230
        type generic, total addresses 6, allocated 1 (16%), misses 0

Total doors: 0
Appl doors: 0
Normal doors: 0
Queued Packets: 0
```

c.   Display NATs on the Gateway router.

```
Gateway# show ip nat translations
Pro Inside global     Inside local      Outside local     Outside global
icmp 209.165.200.225:0 192.168.1.20:1    192.31.7.1:1      192.31.7.1:0
icmp 209.165.200.225:1 192.168.1.21:1    192.31.7.1:1      192.31.7.1:1
icmp 209.165.200.225:2 192.168.1.22:1    192.31.7.1:1      192.31.7.1:2
```

Note: Depending on how much time has elapsed since you performed the pings from each PC, you may not see all three translations. ICMP translations have a short timeout value.

How many Inside local IP addresses are listed in the sample output above? _____

How many Inside global IP addresses are listed? _____

How many port numbers are used paired with the Inside global addresses _____

What would be the result of pinging the Inside local address of PC-A from the ISP router? Why?

_____

_____

Part 3:   Configure and Verify PAT

In Part 3, you will configure PAT by using an interface instead of a pool of addresses to define the outside address. Not all of the commands in Part 2 will be reused in Part 3.

Step 1:   Clear NATs and statistics on the Gateway router.

Step 2:   Verify the configuration for NAT.

a. Verify that statistics have been cleared.

b. Verify that the outside and inside interfaces are configured for NATs.

c. Verify that the ACL is still configured for NATs.

What command did you use to confirm the results from steps a to c?

_____

Step 3:   Remove the pool of useable public IP addresses.

```
Gateway(config)# no ip nat pool public_access 209.165.200.225 209.165.200.230
netmask 255.255.255.248
```

Step 4:   Remove the NAT translation from inside source list to outside pool.

```
Gateway(config)# no ip nat inside source list 1 pool public_access overload
```

Step 5:   Associate the source list with the outside interface.

```
Gateway(config)# ip nat inside source list 1 interface serial 0/0/1 overload
```

Step 6:   Test the PAT configuration.

a. From each PC, ping the 192.31.7.1 address on the ISP router.

b. Display NAT statistics on the Gateway router.

```
Gateway# show ip nat statistics
Total active translations: 3 (0 static, 3 dynamic; 3 extended)
Peak translations: 3, occurred 00:00:19 ago
Outside interfaces:
  Serial0/0/1
Inside interfaces:
  GigabitEthernet0/1
Hits: 24  Misses: 0
CEF Translated packets: 24, CEF Punted packets: 0
Expired translations: 0
Dynamic mappings:
-- Inside Source
[Id: 2] access-list 1 interface Serial0/0/1 refcount 3

Total doors: 0
Appl doors: 0
Normal doors: 0
Queued Packets: 0
```

c. Display NAT translations on Gateway.

```
Gateway# show ip nat translations
Pro Inside global      Inside local      Outside local      Outside global
icmp 209.165.201.18:3  192.168.1.20:1    192.31.7.1:1       192.31.7.1:3
icmp 209.165.201.18:1  192.168.1.21:1    192.31.7.1:1       192.31.7.1:1
icmp 209.165.201.18:4  192.168.1.22:1    192.31.7.1:1       192.31.7.1:4
```

Reflection

What advantages does PAT provide?

_____

Router Interface Summary Table

Router Interface Summary				
Router Model	Ethernet Interface #1	Ethernet Interface #2	Serial Interface #1	Serial Interface #2
1800	Fast Ethernet 0/0 (F0/0)	Fast Ethernet 0/1 (F0/1)	Serial 0/0/0 (S0/0/0)	Serial 0/0/1 (S0/0/1)
1900	Gigabit Ethernet 0/0 (G0/0)	Gigabit Ethernet 0/1 (G0/1)	Serial 0/0/0 (S0/0/0)	Serial 0/0/1 (S0/0/1)
2801	Fast Ethernet 0/0 (F0/0)	Fast Ethernet 0/1 (F0/1)	Serial 0/1/0 (S0/1/0)	Serial 0/1/1 (S0/1/1)
2811	Fast Ethernet 0/0 (F0/0)	Fast Ethernet 0/1 (F0/1)	Serial 0/0/0 (S0/0/0)	Serial 0/0/1 (S0/0/1)
2900	Gigabit Ethernet 0/0 (G0/0)	Gigabit Ethernet 0/1 (G0/1)	Serial 0/0/0 (S0/0/0)	Serial 0/0/1 (S0/0/1)
Note: To find out how the router is configured, look at the interfaces to identify the type of router and how many interfaces the router has. There is no way to effectively list all the combinations of configurations for each router class. This table includes identifiers for the possible combinations of Ethernet and Serial interfaces in the device. The table does not include any other type of interface, even though a specific router may contain one. An example of this might be an ISDN BRI interface. The string in parenthesis is the legal abbreviation that can be used in Cisco IOS commands to represent the interface.				

11.3.1.5 Lab - Troubleshooting NAT Configurations

Topology

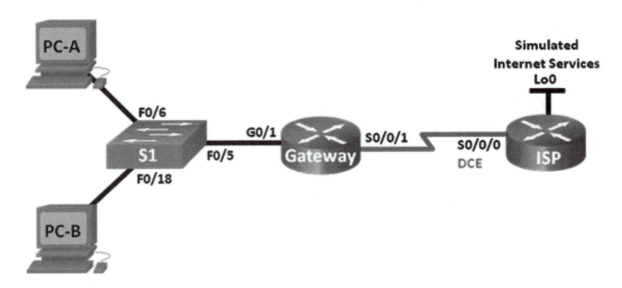

Addressing Table

Device	Interface	IP Address	Subnet Mask	Default Gateway
Gateway	G0/1	192.168.1.1	255.255.255.0	N/A
	S0/0/1	209.165.200.225	255.255.255.252	N/A
ISP	S0/0/0 (DCE)	209.165.200.226	255.255.255.252	N/A
	Lo0	198.133.219.1	255.255.255.255	N/A
PC-A	NIC	192.168.1.3	255.255.255.0	192.168.1.1
PC-B	NIC	192.168.1.4	255.255.255.0	192.168.1.1

Objectives

Part 1: Build the Network and Configure Basic Device Settings

Part 2: Troubleshoot Static NAT

Part 3: Troubleshoot Dynamic NAT

Background / Scenario

In this lab, the Gateway router was configured by an inexperienced network administrator at your company. Several errors in the configuration have resulted in NAT issues. Your boss has asked you to troubleshoot and correct the NAT errors and document your work. Ensure that the network supports the following:

- PC-A acts as a web server with a static NAT and will be reachable from the outside using the 209.165.200.254 address.

- PC-B acts as a host computer and dynamically receives an IP address from the created pool of addresses called NAT_POOL, which uses the 209.165.200.240/29 range.

Note: The routers used with CCNA hands-on labs are Cisco 1941 Integrated Services Routers (ISRs) with Cisco IOS Release 15.2(4)M3 (universalk9 image). The switches used are Cisco Catalyst 2960s with Cisco IOS Release 15.0(2) (lanbasek9 image). Other routers, switches and Cisco IOS versions can be used. Depending on the model and Cisco IOS version, the commands available and output produced might vary from what is shown in the labs. Refer to the Router Interface Summary Table at the end of this lab for the correct interface identifiers.

Note: Make sure that the routers and switch have been erased and have no startup configurations. If you are unsure, contact your instructor.

Required Resources

- 2 Routers (Cisco 1941 with Cisco IOS Release 15.2(4)M3 universal image or comparable)
- 1 Switch (Cisco 2960 with Cisco IOS Release 15.0(2) lanbasek9 image or comparable)
- 2 PCs (Windows 7, Vista, or XP with terminal emulation program, such as Tera Term)
- Console cables to configure the Cisco IOS devices via the console ports
- Ethernet and serial cables as shown in the topology

Part 1: Build the Network and Configure Basic Device Settings

In Part 1, you will set up the network topology and configure the routers with basic settings. Additional NAT-related configurations are provided. The NAT configurations for the Gateway router contains errors that you will identify and correct as you proceed through the lab.

Step 1: Cable the network as shown in the topology.

Step 2: Configure PC hosts.

Step 3: Initialize and reload the switch and routers.

Step 4: Configure basic settings for each router.

a. Disable DNS lookup.

b. Configure device name as shown in the topology.

c. Configure IP addresses as listed in the Address Table.

d. Set the clock rate to **128000** for DCE serial interfaces.

e. Assign **cisco** as the console and vty password.

f. Assign **class** as the encrypted privileged EXEC mode password.

g. Configure **logging synchronous** to prevent console messages from interrupting the command entry.

Step 5: Configure static routing.

a. Create a static route from the ISP router to the Gateway router-assigned public network address range 209.165.200.224/27.

```
ISP(config)# ip route 209.165.200.224 255.255.255.224 s0/0/0
```

b.   Create a default route from the Gateway router to the ISP router.

```
Gateway(config)# ip route 0.0.0.0 0.0.0.0 s0/0/1
```

Step 6:   Load router configurations.

The configurations for the routers are provided for you. There are errors with the configuration for the Gateway router. Identify and correct the configurations errors.

Gateway Router Configuration

```
interface g0/1
 ip nat outside


 no shutdown
interface s0/0/0
 ip nat outside


interface s0/0/1


 no shutdown
ip nat inside source static 192.168.2.3 209.165.200.254


ip nat pool NAT_POOL 209.165.200.241 209.165.200.246 netmask 255.255.255.248
ip nat inside source list NAT_ACL pool NATPOOL


ip access-list standard NAT_ACL
 permit 192.168.10.0 0.0.0.255


banner motd $AUTHORIZED ACCESS ONLY$
end
```

Step 7:   Save the running configuration to the startup configuration.

Part 2:   Troubleshoot Static NAT

In Part 2, you will examine the static NAT for PC-A to determine if it is configured correctly. You will troubleshoot the scenario until the correct static NAT is verified.

a.   To troubleshoot issues with NAT, use the **debug ip nat** command. Turn on NAT debugging to see translations in real-time across the Gateway router.

```
Gateway# debug ip nat
```

b.   From PC-A, ping Lo0 on the ISP router. Do any NAT debug translations appear on the Gateway router?

c. On the Gateway router, enter the command that allows you to see all current NAT translations on the Gateway router. Write the command in the space below.

_____

Why are you seeing a NAT translation in the table, but none occurred when PC-A pinged the ISP loopback interface? What is needed to correct the issue?

_____

d. Record any commands that are necessary to correct the static NAT configuration error.

_____

_____

e. From PC-A, ping Lo0 on the ISP router. Do any NAT debug translations appear on the Gateway router?

_____

f. On the Gateway router, enter the command that allows you to observe the total number of current NATs. Write the command in the space below.

_____

Is the static NAT occurring successfully? Why?

_____

g. On the Gateway router, enter the command that allows you to view the current configuration of the router. Write the command in the space below.

_____

h. Are there any problems with the current configuration that prevent the static NAT from occurring?

_____

i. Record any commands that are necessary to correct the static NAT configuration errors.

_____

_____

_____

j. From PC-A, ping Lo0 on the ISP router. Do any NAT debug translations appear on the Gateway router?

_____

k. Use the **show ip nat translations verbose** command to verify static NAT functionality.

   Note: The timeout value for ICMP is very short. If you do not see all the translations in the output, redo the ping.

   Is the static NAT translation occurring successfully? _____

   If static NAT is not occurring, repeat the steps above to troubleshoot the configuration.

Part 3: **Troubleshoot Dynamic NAT**

a. From PC-B, ping Lo0 on the ISP router. Do any NAT debug translations appear on the Gateway router?

   _____

b. On the Gateway router, enter the command that allows you to view the current configuration of the router. Are there any problems with the current configuration that prevent dynamic NAT from occurring?

_____

c. Record any commands that are necessary to correct the dynamic NAT configuration errors.

_____

_____

_____

d. From PC-B, ping Lo0 on the ISP router. Do any NAT debug translations appear on the Gateway router?

_____

e. Use the **show ip nat statistics** to view NAT usage.

Is the NAT occurring successfully? _____

What percentage of dynamic addresses has been allocated? _____

f. Turn off all debugging using the **undebug all** command.

Reflection

1. What is the benefit of a static NAT?

2. What issues would arise if 10 host computers in this network were attempting simultaneous Internet communication?

Router Interface Summary Table

Router Interface Summary				
Router Model	**Ethernet Interface #1**	**Ethernet Interface #2**	**Serial Interface #1**	**Serial Interface #2**
1800	Fast Ethernet 0/0 (F0/0)	Fast Ethernet 0/1 (F0/1)	Serial 0/0/0 (S0/0/0)	Serial 0/0/1 (S0/0/1)
1900	Gigabit Ethernet 0/0 (G0/0)	Gigabit Ethernet 0/1 (G0/1)	Serial 0/0/0 (S0/0/0)	Serial 0/0/1 (S0/0/1)
2801	Fast Ethernet 0/0 (F0/0)	Fast Ethernet 0/1 (F0/1)	Serial 0/1/0 (S0/1/0)	Serial 0/1/1 (S0/1/1)
2811	Fast Ethernet 0/0 (F0/0)	Fast Ethernet 0/1 (F0/1)	Serial 0/0/0 (S0/0/0)	Serial 0/0/1 (S0/0/1)
2900	Gigabit Ethernet 0/0 (G0/0)	Gigabit Ethernet 0/1 (G0/1)	Serial 0/0/0 (S0/0/0)	Serial 0/0/1 (S0/0/1)
Note: To find out how the router is configured, look at the interfaces to identify the type of router and how many interfaces the router has. There is no way to effectively list all the combinations of configurations for each router class. This table includes identifiers for the possible combinations of Ethernet and Serial interfaces in the device. The table does not include any other type of interface, even though a specific router may contain one. An example of this might be an ISDN BRI interface. The string in parenthesis is the legal abbreviation that can be used in Cisco IOS commands to represent the interface.				

11.4.1.1 Class Activity – NAT Check

Objective

Configure, verify and analyze static NAT, dynamic NAT and NAT with overloading.

Scenario

Network address translation is not currently included in your company's network design. It has been decided to configure some devices to use NAT services for connecting to the mail server.

Before deploying NAT live on the network, you prototype it using a network simulation program.

Resources

- Packet Tracer software
- Word processing or presentation software

Directions

Step 1:   **Create a very small network topology using Packet Tracer, including, at minimum:**

 a.   Two 1941 routers, interconnected

 b.   Two LAN switches, one per router

 c.   One mail server, connected to the LAN on one router

 d.   One PC or laptop, connected the LAN on the other router

Step 2:   **Address the topology.**

 a.   Use private addressing for all networks, hosts, and device.

 b.   DHCP addressing of the PC or laptop is optional.

 c.   Static addressing of the mail server is mandatory.

Step 3:   **Configure a routing protocol for the network.**

Step 4:   **Validate full network connectivity without NAT services.**

 a.   Ping from one end of the topology and back to ensure the network is functioning fully.

 b.   Troubleshoot and correct any problems preventing full network functionality.

Step 5:   **Configure NAT services on either router from the host PC or laptop to the mail server**

Step 6:   **Produce output validating NAT operations on the simulated network.**

 a.   Use the **show ip nat statistics**, **show access-lists**, and **show ip nat translations** commands to gather information about NAT's operation on the router

 b.   Copy and paste or save screenshots of the topology and output information to a word processing or presentation document.

Step 7:   **Explain the NAT design and output to another group or to the class.**

Appendix A — Supplemental Labs

0.0.0.1 Lab – Initializing and Reloading a Router and Switch

Topology

Objectives

Part 1: Set Up Devices in the Network as Shown in the Topology

Part 2: Initialize the Router and Reload

Part 3: Initialize the Switch and Reload

Background / Scenario

Before starting a CCNA hands-on lab that makes use of either a Cisco router or switch, ensure that the devices in use have been erased and have no startup configurations present. Otherwise, the results of your lab may be unpredictable. This lab provides a detail procedure for initializing and reloading a Cisco router and a Cisco switch.

Note: The routers used with CCNA hands-on labs are Cisco 1941 Integrated Services Routers (ISRs) with Cisco IOS Release 15.2(4)M3 (universalk9 image). The switches used are Cisco Catalyst 2960s with Cisco IOS Release 15.0(2) (lanbasek9 image). Other routers, switches, and Cisco IOS versions can be used. Depending on the model and Cisco IOS version, the commands available and output produced might vary from what is shown in the labs.

Required Resources

- 1 Router (Cisco 1941 with Cisco IOS software, Release 15.2(4)M3 universal image or comparable)
- 1 Switch (Cisco 2960 with Cisco IOS Release 15.0(2) lanbasek9 image or comparable)
- 2 PCs (Windows 7, Vista, or XP with terminal emulation program, such as Tera Term)
- Console cables to configure the Cisco IOS devices via the console ports

Part 1: Set Up Devices in the Network as Shown in the Topology

Step 1: Cable the network as shown in the topology.

Attach console cables to the devices shown in the topology diagram.

Step 2: Power on all the devices in the topology.

Wait for all devices to finish the software load process before moving to Part 2.

Part 2: Initialize the Router and Reload

Step 1: Connect to the router.

Console into the router and enter privileged EXEC mode using the **enable** command.

```
Router> enable
Router#
```

Step 2: Erase the startup configuration file from NVRAM.

Type the **erase startup-config** command to remove the startup configuration from nonvolatile random-access memory (NVRAM).

```
Router# erase startup-config
Erasing the nvram filesystem will remove all configuration files! Continue? [confirm]
[OK]
Erase of nvram: complete
Router#
```

Step 3: Reload the router.

Issue the **reload** command to remove an old configuration from memory. When prompted to Proceed with reload, press Enter to confirm the reload. Pressing any other key will abort the reload.

```
Router# reload
Proceed with reload? [confirm]


*Nov 29 18:28:09.923: %SYS-5-RELOAD: Reload requested by console. Reload Reason:
Reload Command.
```

Note: You may receive a prompt to save the running configuration prior to reloading the router. Respond by typing **no** and press Enter.

```
System configuration has been modified. Save? [yes/no]: no
```

Step 4: Bypass the initial configuration dialog.

After the router reloads, you are prompted to enter the initial configuration dialog. Enter **no** and press Enter.

```
Would you like to enter the initial configuration dialog? [yes/no]: no
```

Step 5:   **Terminate the autoinstall program.**

You will be prompted to terminate the autoinstall program. Respond **yes** and then press Enter.

```
Would you like to terminate autoinstall? [yes]: yes
Router>
```

Part 3:   **Initialize the Switch and Reload**

Step 1:   **Connect to the switch.**

Console into the switch and enter privileged EXEC mode.

```
Switch> enable
Switch#
```

Step 2:   **Determine if there have been any virtual local-area networks (VLANs) created.**

Use the **show flash** command to determine if any VLANs have been created on the switch.

```
Switch# show flash


Directory of flash:/


    2  -rwx        1919   Mar 1 1993 00:06:33 +00:00   private-config.text
    3  -rwx        1632   Mar 1 1993 00:06:33 +00:00   config.text
    4  -rwx       13336   Mar 1 1993 00:06:33 +00:00   multiple-fs
    5  -rwx    11607161   Mar 1 1993 02:37:06 +00:00   c2960-lanbasek9-mz.150-2.SE.bin
    6  -rwx         616   Mar 1 1993 00:07:13 +00:00   vlan.dat


32514048 bytes total (20886528 bytes free)
Switch#
```

Step 3:   **Delete the VLAN file.**

a.   If the **vlan.dat** file was found in flash, then delete this file.

```
Switch# delete vlan.dat
Delete filename [vlan.dat]?
```

You will be prompted to verify the file name. At this point, you can change the file name or just press Enter if you have entered the name correctly.

b.   When you are prompted to delete this file, press Enter to confirm the deletion. (Pressing any other key will abort the deletion.)

```
Delete flash:/vlan.dat? [confirm]
Switch#
```

Step 4:   **Erase the startup configuration file.**

Use the **erase startup-config** command to erase the startup configuration file from NVRAM. When you are prompted to remove the configuration file, press Enter to confirm the erase. (Pressing any other key will abort the operation.)

```
Switch# erase startup-config
Erasing the nvram filesystem will remove all configuration files! Continue? [confirm]
[OK]
Erase of nvram: complete
Switch#
```

Step 5:   **Reload the switch.**

Reload the switch to remove any old configuration information from memory. When you are prompted to re-load the switch, press Enter to proceed with the reload. (Pressing any other key will abort the reload.)

```
Switch# reload
Proceed with reload? [confirm]
```

Note: You may receive a prompt to save the running configuration prior to reloading the switch. Type **no** and press Enter.

```
System configuration has been modified. Save? [yes/no]: no
```

Step 6:   **Bypass the initial configuration dialog.**

After the switch reloads, you should see a prompt to enter the initial configuration dialog. Type **no** at the prompt and press Enter.

```
Would you like to enter the initial configuration dialog? [yes/no]: no
Switch>
```

Reflection

1. Why is it necessary to erase the startup configuration before reloading the router?

2. You find a couple configurations issues after saving the running configuration to the startup configuration, so you make the necessary changes to fix those issues. If you were to reload the device now, what configuration would be restored to the device after the reload?

0.0.0.2 Lab – Installing the IPv6 Protocol and Assigning Host Addresses with Windows XP

Objectives

Part 1: Install the IPv6 Protocol on a Windows XP PC

- Install the IPv6 protocol.
- Examine IPv6 address information.

Part 2: Use the Network Shell (netsh) Utility

- Work inside the **netsh** utility.
- Configure a static IPv6 address on the local-area network (LAN) interface.
- Exit the **netsh** utility.
- Display IPv6 address information using **netsh**.
- Issue **netsh** instructions from the command prompt.

Background / Scenario

The Internet Protocol Version 6 (IPv6) is not enabled by default in Windows XP. Windows XP includes IPv6 implementation, but the IPv6 protocol must be installed. XP does not provide a way to configure IPv6 static addresses from the Graphical User Interface (GUI), so all IPv6 static address assignments must be done using the Network Shell (**netsh**) utility.

In this lab, you will install the IPv6 protocol on a Windows XP PC. You will then assign a static IPv6 address to the LAN interface.

Required Resources

1 Windows XP PC

Part 1: Install the IPv6 Protocol on a Windows XP PC

In Part 1, you will install the IPv6 protocol on a PC running Windows XP. You will also use two commands to view the IPv6 addresses assigned to the PC.

Step 1: Install the IPv6 protocol.

From the command prompt window, type **ipv6 install** to install the IPv6 protocol.

Step 2: **Examine IPv6 Address Information.**

Use the **ipconfig /all** command to view IPv6 address information.

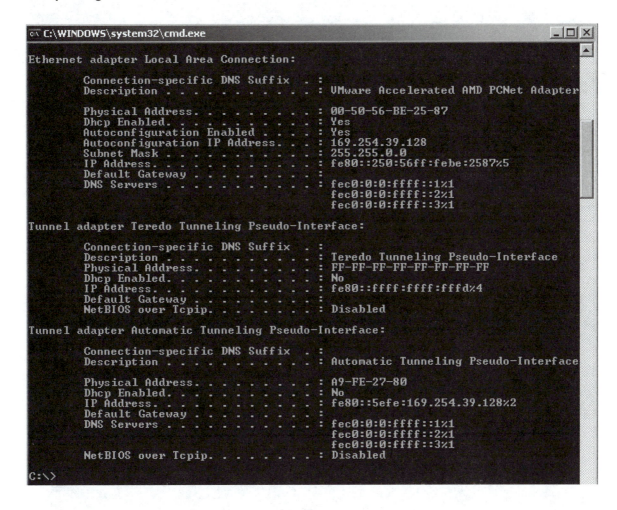

Part 2: **Use the Network Shell (netsh) Utility**

Network Shell (**netsh**) is a command-line utility included with Windows XP and newer Windows operating systems, such as Vista and Windows 7. It allows you to configure the IPv6 address information on your LAN. In Part 2, you will use the **netsh** utility to configure static IPv6 address information on a Windows XP PC LAN interface. You will also use the **netsh** utility to display the PC LAN interface IPv6 address information.

Step 1: **Work inside the Network Shell utility.**

a. From the command prompt window, type **netsh** and press Enter to start the **netsh** utility. The command prompt changes from **C:\>** to **netsh>**.

b. At the prompt, enter a question mark (**?**) and press Enter to provide the list of available parameters.

```
netsh>?

The following commands are available:

Commands in this context:
..              - Goes up one context level.
?               - Displays a list of commands.
abort           - Discards changes made while in offline mode.
add             - Adds a configuration entry to a list of entries.
alias           - Adds an alias.
bridge          - Changes to the `netsh bridge' context.
bye             - Exits the program.
commit          - Commits changes made while in offline mode.
delete          - Deletes a configuration entry from a list of entries.
diag            - Changes to the `netsh diag' context.
dump            - Displays a configuration script.
exec            - Runs a script file.
exit            - Exits the program.
firewall        - Changes to the `netsh firewall' context.
help            - Displays a list of commands.
interface       - Changes to the `netsh interface' context.
lan             - Changes to the `netsh lan' context.
nap             - Changes to the `netsh nap' context.
offline         - Sets the current mode to offline.
online          - Sets the current mode to online.
popd            - Pops a context from the stack.
pushd           - Pushes current context on stack.
quit            - Exits the program.
ras             - Changes to the `netsh ras' context.
routing         - Changes to the `netsh routing' context.
set             - Updates configuration settings.
show            - Displays information.
unalias         - Deletes an alias.
winsock         - Changes to the `netsh winsock' context.

The following sub-contexts are available:
 bridge diag firewall interface lan nap ras routing winsock

To view help for a command, type the command, followed by a space, and then
 type ?.

netsh>
```

c.   Type **interface ?** and press Enter to provide the list of interface commands.

```
netsh>interface ?

The following commands are available:

Commands in this context:
?               - Displays a list of commands.
add             - Adds a configuration entry to a table.
delete          - Deletes a configuration entry from a table.
dump            - Displays a configuration script.
help            - Displays a list of commands.
ip              - Changes to the `netsh interface ip' context.
ipv6            - Changes to the `netsh interface ipv6' context.
portproxy       - Changes to the `netsh interface portproxy' context.
reset           - Resets information.
set             - Sets configuration information.
show            - Displays information.

The following sub-contexts are available:
 ip ipv6 portproxy

To view help for a command, type the command, followed by a space, and then
 type ?.

netsh>
```

Note: You can use the question mark (**?**) at any level in the **netsh** utility to list the available options. The up arrow can be used to scroll through previous **netsh** commands. The **netsh** utility also allows you to abbreviate commands, as long as the abbreviation is unique.

Step 2: **Configure a static IPv6 address on the LAN interface.**

To add a static IPv6 address to the LAN interface, issue the **interface ipv6 add address** command from inside the **netsh** utility.

```
netsh>interface ipv6 add address "Local Area Connection" 2001:db8:acad:a::3
Ok.

netsh>
```

Step 3: **Display IPv6 address information using the netsh utility.**

You can display IPv6 address information using the **interface ipv6 show address** command.

```
netsh>interface ipv6 show address
Querying active state...

Interface 5: Local Area Connection

Addr Type  DAD State  Valid Life    Pref. Life    Address
---------  ---------  -----------   -----------   -------
Manual     Preferred    infinite      infinite 2001:db8:acad:a::3
Link       Preferred    infinite      infinite fe80::250:56ff:febe:2587

Interface 4: Teredo Tunneling Pseudo-Interface

Addr Type  DAD State  Valid Life    Pref. Life    Address
---------  ---------  -----------   -----------   -------
Link       Preferred    infinite      infinite fe80::ffff:ffff:fffd

Interface 2: Automatic Tunneling Pseudo-Interface

Addr Type  DAD State  Valid Life    Pref. Life    Address
---------  ---------  -----------   -----------   -------
Link       Preferred    infinite      infinite fe80::5efe:169.254.39.128

Interface 1: Loopback Pseudo-Interface

Addr Type  DAD State  Valid Life    Pref. Life    Address
---------  ---------  -----------   -----------   -------
Loopback   Preferred    infinite      infinite ::1
Link       Preferred    infinite      infinite fe80::1

netsh>
```

Step 4: **Exit the netsh utility.**

Use the **exit** command to exit from the **netsh** utility.

```
netsh>exit

C:\>
```

Step 5: **Issue netsh instructions from the command prompt.**

All **netsh** instructions can be entered from the command prompt, outside the **netsh** utility, by preceding the instruction with the **netsh** command.

```
C:\>netsh interface ipv6 show address
Querying active state...

Interface 5: Local Area Connection

Addr Type  DAD State   Valid Life   Pref. Life   Address
---------  ---------   ----------   ----------   -------
Manual     Preferred     infinite     infinite 2001:db8:acad:a::3
Link       Preferred     infinite     infinite fe80::250:56ff:febe:2587

Interface 4: Teredo Tunneling Pseudo-Interface

Addr Type  DAD State   Valid Life   Pref. Life   Address
---------  ---------   ----------   ----------   -------
Link       Preferred     infinite     infinite fe80::ffff:ffff:fffd

Interface 2: Automatic Tunneling Pseudo-Interface

Addr Type  DAD State   Valid Life   Pref. Life   Address
---------  ---------   ----------   ----------   -------
Link       Preferred     infinite     infinite fe80::5efe:169.254.39.128

Interface 1: Loopback Pseudo-Interface

Addr Type  DAD State   Valid Life   Pref. Life   Address
---------  ---------   ----------   ----------   -------
Loopback   Preferred     infinite     infinite ::1
Link       Preferred     infinite     infinite fe80::1

C:\>
```

Reflection

1. How would you renew your LAN interface address information from the **netsh** utility?

   Hint: Use the question mark (**?**) for help in obtaining the parameter sequence.

   _____

   _____